MW00638404

The Darwinian Trap

The Darwinian Trap

THE HIDDEN EVOLUTIONARY FORCES THAT EXPLAIN OUR WORLD (AND THREATEN OUR FUTURE)

Kristian Rönn

CROWN CURRENCY
NEW YORK

CROWN CURRENCY
An imprint of the Crown Publishing Group
A division of Penguin Random House LLC.

crownpublishing.com

Illustrations by Veronica Rönn

Library of Congress Cataloging-in-Publication Data
Names: Rönn, Kristian, author.
Title: The Darwinian trap : the hidden evolutionary forces that explain our world (and threaten our future) / Kristian Rönn.
Identifiers: LCCN 2024007297 (print) | LCCN 2024007298 (ebook) | ISBN 9780593594056 (hardcover) | ISBN 9780593594070 (trade paperback) | ISBN 9780593594063 (ebook)
Subjects: LCSH: Social Darwinism. | Social evolution. | Social values.
Classification: LCC HM631 .R66 2024 (print) | LCC HM631 (ebook) | DDC 303.4—dc23/eng/20240221
LC record available at https://lccn.loc.gov/2024007297
LC ebook record available at https://lccn.loc.gov/2024007298

ISBN: 978-0-593-59405-6
Ebook ISBN: 978-0-593-59406-3

Printed in the United States of America on acid-free paper

Editor: Paul Whitlatch
Editorial assistant: Katie Berry
Production editor: Terry Deal
Production editorial assistant: Taylor Teague
Text designer: Aubrey Khan
Production manager: Dustin Amick
Managing editors: Christine Tanigawa and Liza Stepanovich
Copy editor: Janet Biehl
Proofreader: JoAnna Kremer
Indexer: Elise Hess
Publicists: Stacey Stein and Dyana Messina
Marketers: Chantelle Walker and Julie Cepler

9 8 7 6 5 4 3 2 1

First Edition

We have Paleolithic emotions, medieval institutions, and god-like technology.

—E. O. WILSON

CONTENTS

PART III

How to Save Life from Life Itself

PROLOGUE

The Parable of Picher

At the dawn of the twenty-first century, Picher, Oklahoma, was arguably the unluckiest town in America. For decades its population had steadily declined—and not just because of the lack of jobs there. Picher's infant mortality rate was unusually high. The town was pockmarked with sinkholes. And the people of Picher, young and old alike, seemed uniquely prone to serious illness: heart disease, lung disease, cancer, and many other ailments. Every town nearby had suffered since the demise of the mining industry there. But Picher seemed particularly cursed.

The town's curse had once been its lifeblood. Picher sits in the far northeastern corner of Oklahoma, at the edge of the Ozark Plateau, near the triple border with Kansas and Missouri, in an area known as the Tri-State Region. While the region might strike contemporary travelers as desolate, a century ago it boasted an immense commercial advantage: vast mineral deposits buried deep underground. In the 1920s, the Tri-State Region mined more zinc and lead than anywhere else in the world, at a time when those minerals were critical to the production of paint, pipes, brass, housewares, armaments, and many other important goods.

The miners of Picher had a nickname for the zinc and lead that abounded there: "King Jack." For decades, Jack was a generous monarch. The Tri-State lead and zinc mines paid high wages, allowing the miners to build relative wealth while the surrounding region flourished. "There is probably no town in the world that has changed so much in the last year or two as Picher," a local newspaper boasted in 1920. New structures were rising every day: shops, cinemas, a bowling alley—and, increasingly, the mounds of accumulated mining detritus known as chat piles.

A miner's output went directly to a mill, where pressure was applied to extract the valuable metals from the surrounding ore. The remaining ore and rock were reduced to a particulate substance known as mine tailings, or "chat." In Picher, those tailings were dumped in heaping piles on the outskirts of town. These chat piles soon overshadowed any of the man-made structures in Picher, with some rising as high as two hundred feet. When the wind blew, it carried a fine mist of pulverized ore and silica dust that drifted through people's lives and lungs.

The boom didn't last—it never does—and the work dried up as the mineral deposits dwindled. By the 1960s, the Picher mines had been picked clean, leaving behind a declining population of unemployed laborers, mine widows, and the mountainous chat piles whose presence evoked better days. As time passed, doctors noticed an alarming rate of other ailments among those who remained in and around Picher: blood diseases, and cognitive impairments, and various types of cancer.

In his 1940 film *Men and Dust,* the photographer Sheldon Dick had depicted the high incidence of debilitating illness among the region's residents. "Silicosis, tuberculosis, malnutrition, general diseases. In the meantime, they sicken. In the meantime, they die," the narrator intones as the camera lingers on a shot of a makeshift graveyard. "Give us health, give us work, give us life," he says at the end of the brief documentary, his voice soon rising to a plaintive cry. "Give us health! Give us work! Give us life!"

But the people of Picher would get none of those things. More than half a century after Dick's documentary, studies showed that locals contracted the lung disease pneumoconiosis at a rate 2,000 percent higher than the national average. In the 1990s, another study showed that 38.3 percent of the children in Picher had elevated blood lead levels, compared with 2 percent of children in the rest of the state.

The key to these diagnoses was readily apparent. You could find it among the dead fish in the local creek, which had grown toxic as soon as the abandoned mines filled up with groundwater; in the swimming holes where children would play on hot summer days, sometimes to emerge with their hair turned orange from the acidic waters; and yes, in the massive chat piles, which had carried metallic dust and trace amounts of lead into locals' bloodstreams for decades. In the early 1980s, the Environmental Protection Agency proclaimed the region one of the most contaminated areas in the United States. By the 2000s, Picher and its environs were so toxic that state and federal agencies offered cash buyouts to encourage its few remaining residents to leave.

Today Picher is a bona fide ghost town, its once-lively streets empty except for rotting buildings and the dust blowing off the chat piles. Lead and zinc put Picher, Oklahoma, on the map. One hundred years later, they have erased it from the face of the earth.

The Wrong Way to Kill a Rat

Why did things go so wrong in Picher? How did an industrial process meant to meet the world's needs end up destroying an entire region? It is tempting to blame the era's mining executives for being so heedless of worker safety and environmental health. But cathartic though it may be to portray the Picher bigwigs as sneering Dickensian villains, it is unlikely that they intended to give their employees cancer and leave the town a poisonous wasteland. Their real goal, presumably, was just to capitalize on the bull market for lead and zinc by getting those minerals out of the ground as fast as possible.

Instead of interpreting the story of Picher as a simple tale of corporate greed, I view it as a parable that illuminates a much broader point. When people or groups choose to optimize around narrow goals—such as quickly and efficiently mining valuable minerals—they often end up creating unintended negative consequences for other people and the broader environment. This phenomenon is not limited to the minerals industry, or to the world of business. The same dynamic occurs in biology, psychology, economics, politics, human resources, sustainable development, virology, artificial intelligence, and other fields. You have almost certainly sensed it in your workplace, in the dating pool, and in other familiar settings.

Stories of optimization gone wrong recur throughout the history of biological life and human societies. From the smallest cancer cells replicating at the expense of the body, to the largest civilizations pursuing ghoulish nuclear arms races that threaten to destroy all that those civilizations have built, simple and complex systems alike are routinely subverted by myopic success strategies that create more problems than they solve. When people and organizations focus intently on achieving one specific goal, they can inadvertently thwart the pursuit of more holistically meaningful outcomes such as well-being, safety, or prosperity. In other words, when we feel compelled to pursue narrow success metrics, and when these metrics become detached from the values that we actually care about, then, as in Picher, we can lose sight of the big picture and bring about widespread harmful consequences.

These consequences aren't always as cataclysmic as they were in Picher. Take, for example, the darkly comic story of the Great Hanoi Rat Massacre. In 1902 the Vietnamese capital had a rat problem, and French colonial officials had a bright idea for how to solve it: they would offer a cash bounty for every rat tail that residents brought to city officials. The French surely assumed that the prospect of a per-tail cash reward would inspire entrepreneurial exterminators to enter the city's sewers and make their fortunes, one dead rat at a time, until the rat problem was under control. In practice, things worked out very differently.

The rat catchers of Hanoi soon realized the limits of the bounty program. If they succeeded in eliminating the rats, then the program would end, and they would lose their meal ticket. So instead of killing the rats, these clever souls began cutting off their tails and setting them free to go procreate and produce more vermin. As the program continued, the French were surprised to find that the city's rat population continued to flourish. Some entrepreneurs went so far as to start breeding rats at home; others set up full-fledged rat farms out in the countryside. Within months of its launch, the beleaguered French canceled the bounty program entirely. *Sacré bleu!*

As with Picher, it is tempting to interpret the failure of the tail-bounty initiative as a matter of individual greed and dishonesty. If only the rat catchers had followed the "spirit" of the program, then maybe the program would have worked! While there is some merit to this analysis, I believe it is ultimately more instructive to understand the many ways in which the program itself promoted devious strategies. In this case, the outcome that the colonial government wanted to achieve—health and hygiene—wasn't fully captured by the measurement they used to track their progress toward that outcome: rat tails collected.

The story of the Great Hanoi Rat Massacre is a classic example of what economists refer to as perverse incentives, or incentives that inadvertently exacerbate the initial problem. Perverse incentives encourage people to game systems for their own benefit, even if doing so might make things worse for everyone else. But I believe that economic theory alone cannot sufficiently explain the underlying impulse to pursue short-term advantage at the expense of broader group or societal welfare. This self-serving impulse is not exclusive to humanity. Indeed, it has existed throughout the history of biological life. It is an evolutionary feature—or, you might say, a glitch—embedded deep within the process of natural selection.

I've given this phenomenon a name: Darwinian demons.

The term *Darwinian demons* refers to pressures—or, more precisely, *selection pressures*—to engage in short-sighted, goal-oriented behaviors that, over time, can yield widespread net-negative consequences. The term *selection pressure* is surely familiar to those of you who have studied evolution. You can think of it as a specific measure of success within a given environment, or as an aspect of an environment that might affect an entity's survival chances therein. A world that is mostly ocean, for example, creates a strong selection pressure for knowing how to swim—or at least for owning a reliable Jet Ski.

Demons, though, is a purposely evocative word, one that you are unlikely to have encountered in your high school biology textbook. The term nevertheless strikes me as apt. Like the archetypal "devil on your shoulder," these evolutionary pressures often act as a pernicious invisible force, encouraging us to make choices that can lead to unwanted outcomes. These demons are *Darwinian* because, as we will explore in Part I of this book, they are directly linked to the process of natural selection, first articulated by Charles Darwin. They arise as a by-product of the adaptation strategies of any agent that is subject to natural selection.

Their biological basis is what renders Darwinian demons distinct from mere "perverse incentives." Unlike those mechanisms, Darwinian demons are not necessarily designed by any specific entity. Instead, they are more elemental; they are intertwined with and inherent to the evolutionary drive for individual survival. This premise in turn raises profound questions about the ways in which we tend to assign blame for bad outcomes—and the steps we might take to most effectively prevent these outcomes.

Armed with our new vocabulary, let's return to the events that transpired in Picher, Oklahoma, and reflect on how Darwinian demons came to devastate that town. The demon at work in Picher was the *selection pressure* for profit—or, more specifically, the economic selection pressure that causes profitable companies to survive and unprofitable ones to fail. The behavior prompted by this demon was high-volume mining paired with low-effort disposal of mine tailings in the form of chat piles. The negative consequences were illness, death, economic decline, and environmental devastation. In the example of the Great Hanoi Rat Massacre, the *selection pressure*—the means by which you could turn a profit in the field of rat catching—was the number of rat tails collected. The resulting behavior was farming rats for their tails, and the negative consequences were the birth and breeding of even more rats that spread diseases.

While the notion of Darwinian demons is my own, it builds on the work of other thinkers. I am an entrepreneur, not an academic. Related ideas have been debated in the ivory tower for many years. And yet these concepts are more than just matters of scholarly abstraction, and their consequences are more than just historical curiosities. In some cases, a Darwinian demon might merely provoke an absurd or unexpected outcome, as we saw with Hanoi's rats. But in the most consequential cases, their effects can potentially imperil the very future of life on Earth. To overcome these challenges, study alone is not enough—entrepreneurial action is required to implement any potential solutions.

Climate change is one such serious threat that we will examine through the lens of Darwinian demons. Nuclear and biological arms races are another. And let's not forget the prospect of runaway artificial intelligence, created and released by companies seeking a first mover's advantage in an unregulated environment. Advanced, unconstrained AI might inflict massive damage on humanity, ranging from widespread unemployment to full-fledged extinction. The scope and impact of these existential risks have been supercharged by our current rate of technological progress. Our inability to adjust our destructive behaviors has put the fate of the whole world at stake.

But if you think that this book plans to make a familiar argument about the excesses and evils of capitalism, you are mistaken. I find it reductive and limiting to view these problems through an exclusively economic lens. My hope is that this book can help shift our perspectives away from the pervasive and simplistic "bad apples" narrative that taints the public dialogue around global issues, and toward more systemic explanations grounded not just in economics but in other fields, too, such as evolutionary biology. The "bad apples" narrative unfairly burdens individual actors with full responsibility for all that is wrong with the world. If we are to have more fruitful discussions about society's biggest problems, then we must adopt a more nuanced narrative about their causes. Instead of merely chastising each flawed apple, we must examine the health of the orchard that so consistently bears these imperfect fruits.

This framework does not seek to absolve individuals of responsibility for harmful actions. To the contrary, humans are ultimately responsible for their choices, and people who make selfish, antisocial ones should be held accountable for doing so. And yet it is important to understand *why* so many people and systems around us seem to be optimizing for the wrong things. My aim is to illuminate the reasons why the success metrics used in corporate and institutional settings so often seem to be out of sync with humanity's core values, such as happiness, health, and well-being. We must forge new, more positive selection pressures if we are to break free from these self-destructive Darwinian demons.

Though my aim is grand, it began with a rather small epiphany. It was in 2014 that I first made these connections, after reading Scott Alexander's enlightening blogpost "Meditations on Moloch." Alexander argues that hidden perverse incentives, which he calls "Moloch" after the Canaanite god of destruction, could be the biggest adversaries to life on our planet—not to mention the root cause of most global suffering. Upon reading this post, I couldn't stop seeing Moloch behind many of our biggest scourges. Climate change caused by the desire to maximize profit; arms races caused by the desire to maximize power; fake news caused by the desire to maximize retention—I realized that all these ills traced back to the same evolutionary phenomenon that leads us to indulge in behaviors that conflict with our collective well-being.

For me, this realization was more than just a fleeting insight: it was a wake-up call. It reminded me that we are all part of an interconnected system, one in which our actions often have a far-reaching impact. Later that year, guided by these ideas, I co-founded a software company called Normative, which is dedicated to recalibrating the societal scales by creating a system of accounting that is better aligned with the things we humans intrinsically value. Our approach has been to start by accounting for greenhouse gas emissions, and then to integrate these methods and principles of "value alignment" into the core incentives of the financial system, much as a profit and loss statement is integrated. My work with Normative has reaffirmed my belief that it is within our collective power to align our actions with our values, for the betterment of all. A society built on selection pressures that promote positive outcomes is not an abstract fantasy—it is an attainable reality.

The first step toward creating that reality is to thoroughly scope the problem we are trying to solve. This book is my attempt to help us all understand the baleful evolutionary mechanisms that, I believe, are the basis for so many social problems, big and small. In it, I will teach you all about Darwinian demons, how they work, and the effects they have had throughout history. My hope is that you will come away from this book with a powerful new lens through which to view our world—both the big-picture problems invoked above, and the more quotidian challenges and frustrations you might encounter in your everyday life. By seeing through the superficial explanations for these problems and refocusing instead on their root causes, we can start to influence the broader system, change the selection pressures, and hence save the world.

■ ■ ■

This book unfolds in three parts. In the opening segment, I will thoroughly introduce you to the concept of Darwinian demons. I'll explore their origins, their presence in our everyday lives, and the powerful influence they have exerted throughout history. I'll also explain why cooperation is the key to neutralizing these demons. At the end of this section, I will take the theory of Darwinian demons a step further and argue that it may very well explain why it seems like we're alone in the universe—and what our apparent galactic solitude might tell us about the inherent fragility of life.

In Part II, I will shift my focus to the present day and identify the most significant Darwinian demons currently threatening our world. By optimizing narrowly around the endless pursuit of resources, power, and intelligence, humans have unleashed a suite of existential risks. The rise of new technologies such as artificial intelligence, the ongoing danger posed by nuclear and biological weapons, the unchecked growth of anthropogenic global warming: all of these forces, if not curtailed, could threaten the future stability of life on Earth, if not the very existence of humanity.

In Part III, I will propose a way forward. The core idea is to replace Darwinian demons with "Darwinian angels," or value-aligned selection pressures that encourage people to promote and safeguard those attributes that humans everywhere genuinely treasure, such as happiness, health, stability, and overall well-being. This section will outline why such a paradigm shift is crucial for the future of our species, and it will offer practical advice for how to counter Darwinian demons and foster a more harmonious future.

The stakes could not be higher. Much as Picher was poisoned into oblivion by the minerals that once provided its prosperity, the world writ large is now threatened by many serious existential risks, born out of humanity's ceaseless pursuit of short-term advantage. Earlier, I referred to the story of Picher as a parable. If we cannot counteract the demons that have brought us to this point, it may well serve as a preview of the future that awaits us.

Our survival within the next century hinges on our ability to confront and conquer these Darwinian demons. But knowledge is always the first step in any battle, and our fight against these formidable foes is no exception. In order to understand how to safeguard the future of the human race, we must first understand the Darwinian demons that seek to throttle our progress.

PART I

An Evolutionary Glitch

CHAPTER ONE

A Crash Course on Darwinian Demons

Charles Darwin could not have fathomed what he was getting into when he boarded the HMS *Beagle* in Plymouth, England, just after Christmas 1831.

A recent graduate of Christ's College, Cambridge, the twenty-two-year-old Darwin had entered school with plans to pursue a career as a country parson. But he also harbored a deep and abiding fascination with the natural world, and he hoped to see as much of it as possible before beginning his adult life. So when the Cambridge botanist John Henslow recommended his former pupil for a naturalist's spot on the *Beagle*'s voyage to conduct a wide-ranging survey of South America—Darwin's wealthy father paid the fare—it didn't take long for Darwin to accept the offer.

This second voyage of the HMS *Beagle*—the first had been marred by the captain's suicide and a hostage crisis—was supposed to last two years. Instead, the journey took five years, with the *Beagle* sailing around Cape Horn, across the Pacific Ocean to Australia and Africa, then back to South America before heading home. The vessel docked in Falmouth on October 2, 1836, by which point Darwin had long since abandoned any thoughts of spending his life in a parsonage. He would soon find very different means of opining on the nature of man.

Darwin's letters home to England over the course of his voyage, recounting all he had seen and experienced during his travels, had already won him a scholar's reputation, one that blossomed into celebrity upon the publication of his first book, *The Voyage of the Beagle,* in 1839. In it, Darwin offered not just a compelling narrative of his long and exciting trip but extensive remarks on the wide variety of life and landscapes he had observed. He was particularly influenced by his experiences on the Galápagos Islands, a remote, sparsely populated archipelago off the coast of Ecuador.

The Galápagos are perhaps best known for their giant tortoises, which can live for hundreds of years. But Darwin's attention was also drawn to the birds there, several specimens of which he brought back to England. Though they looked broadly similar upon first glance, further inspection revealed distinct variations. Some of the birds had long, thin beaks, for example, while others had short, thick ones. Darwin taxonomized the birds separately—labeling some as blackbirds, others as finches, still others as wrens—then basically put them out of his mind until he got home.

But after examining the specimens upon the *Beagle*'s return, the ornithologist John Gould announced that, much to Darwin's surprise, the birds were *all* finches. Moreover, each finch was sufficiently different from its counterparts that Gould classified them as twelve entirely separate species. Reviewing his notes from the voyage, Darwin realized that the birds' variable traits seemed correlated to the differences in their environments. The thick-beaked birds, for example, came from an island where those beaks were ideally suited for cracking the hard-shelled ground nuts that abounded there. These correlations seemed to imply that speciation—the process by which members of a population evolve divergent traits, such that those members eventually become an entirely new species—had occurred in response to those environments.

In the second edition of *The Voyage of the Beagle,* and in much greater depth in 1859's *On the Origin of Species,* Darwin refined his initial observations about the Galápagos finches into a comprehensive theory of evolution, coining the term *natural selection* as a bit of descriptive shorthand. His theory, rendered simply, went something like this: There are always random variations within species. Some people are short, for example, while others are tall, and so on. But if the short and tall individuals within a population happen to share an environment where it is easier to survive as a tall person—an unusual land, perhaps, where predators will not attack any human standing six feet or higher—then the tall people will be more apt to survive. In turn, they will be more likely to reproduce and pass their characteristics down to subsequent generations.

In this hypothetical scenario, we might say that "tall" genes were "naturally selected" for survival by the evolutionary environment. The process by which the possessors of adaptive traits come to thrive in their environments is the essence of natural selection, which is also sometimes referred to as "survival of the fittest." As Darwin put it in his introduction to *On the Origin of Species*:

> As many more individuals of each species are born than can possibly survive; and as, consequently, there is a frequently recurring struggle for existence, it follows that any being, if

> it vary however slightly in any manner profitable to itself, under the complex and sometimes varying conditions of life, will have a better chance of surviving, and thus be naturally selected.

Darwin's theory of evolution via natural selection eventually transformed the natural sciences. By the time of his death in 1882, the scientific community had generally come to accept Darwin's premise that species evolved over time, although it would take several more decades for the theory of natural selection to gain common currency. Eventually, some people began to wonder whether those theories might be more broadly applicable.

In 1983 the evolutionary biologist Richard Dawkins published a paper titled "Universal Darwinism," in which he took Darwin's theories of how life evolved on Earth and suggested that they also could be applied in a wide variety of domains: companies, societies, memes, and so on. Dawkins argued that the process of natural selection would catalyze evolution everywhere, in all qualifying environments: on Earth and in far-flung galaxies, from the finches of South America to the boardrooms of New York.

While Dawkins may have been the first to use the term *universal Darwinism,* he was neither the first nor the last to posit that the principles of natural selection might apply both to biological life and to human structures, processes, and environments. In natural contexts and in social settings, across all known fields of inquiry and endeavor, evolution by natural selection takes place in any system that observes the following characteristics, as first articulated by Darwin:

1. Variation. The system contains a population whose members vary in their characteristics: looks, behaviors, strategies, and so

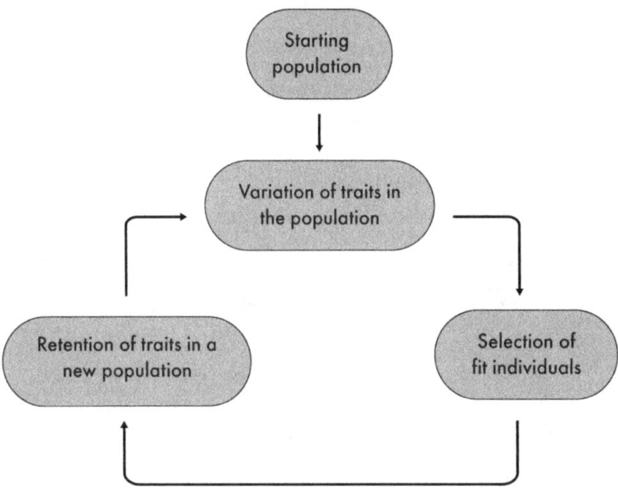

on. Imagine a population of mice, for instance, whose different characteristics include size, speed, and color. Some mice might be small, fast, and white, while others might be large, slow, and gray. For nonbiological populations such as companies, these differing characteristics might be their revenue models. Some digital media companies might pursue a subscription-based model, while others might adopt an advertisement-based model.

2. Selection. Entities with different characteristics have different rates of survival and reproduction, which is sometimes called *differential fitness*. Mice with brown fur might blend into their environment more effectively than their gray-furred counterparts, making them harder to catch for predatory birds flying overhead. When it comes to digital media companies, the market might select companies with an advertisement-based model, since that model attracts more consumers by offering its content for free.

3. Retention. Characteristics that contribute to differential fitness are heritable, and they affect propagation rates in the population. In the previous example, easily camouflaged brown mice would be more likely to survive until they reproduce, which in turn means that their offspring will be more numerous in the general population. In the realm of media companies, the selected company might expand its operations, creating multiple subsidiaries with the same business model characteristics.

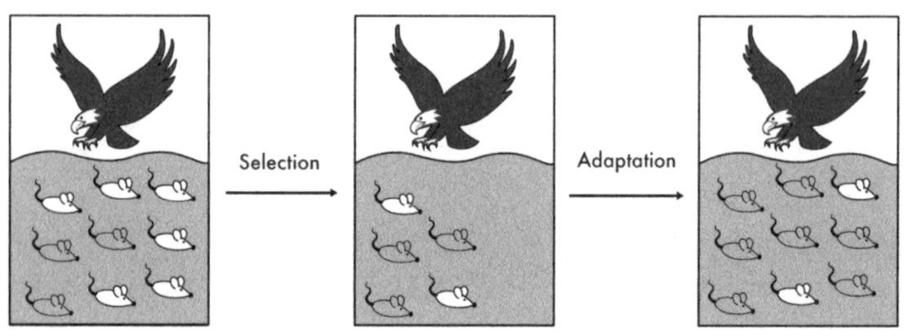

You might contend that a media company is not actually subject to evolutionary dynamics, since its trajectory is steered by human choices rather than by natural selection. But the distinction between "natural" and "unnatural" selection is moot. Regardless of whether we're talking about a bird of prey choosing a mouse or a person choosing a news source, the essence of selection remains unchanged. Evolutionary theory posits that as long as Darwin's principles of variation, selection, and retention are met, adaptive traits will spread, irrespective of the nature of the selection pressures.

In one way or another, we have all had firsthand experience with this process. We live, work, and play in environments where the "winners" are often those best able to adapt to the demands of those environments. Humans are different from mice and other animals, of course. Our highly evolved brains have given us the capacity for self-awareness and empathy; the ability to understand how our actions might affect the future; the power to find virtue in protecting the weakest among us; and other unique attributes. We have used these capacities to devise systems of law and morality that, among other functions, serve to constrain our most primal Darwinian survival instincts. And yet the pressures of our environments consistently lead humans and groups to disregard these "better angels" in the pursuit of narrow individual advantage.

We live in a world in which good ideas can soon turn sour and well-intentioned processes often go awry. Governments and corporations are often led by people who might begin their careers promising positive change, only to end up making self-serving choices meant to maximize their own fitness. In cultural contexts, beauty and truth often get subsumed in a sea of empty, disposable trash. Society is filled with smart, well-meaning people who often truly want to make the world a better place. Why do their efforts so often fail?

I contend that the answer comes down to Darwinian demons, the evolutionary glitch that can lead us to optimize around narrow goals while degrading the broader environment. In this chapter, I'm going to probe the nature of these Darwinian demons, and how they can affect and infect everyday processes and situations. By doing so, I hope to give you new tools with which to interpret and understand the sorts of systemic problems that we too often blame solely on the doings of "bad actors."

While people who do bad things should indeed be held accountable for their actions, the systems that make it profitable for people to do those bad things must be held accountable, too. In the words of Ice Cube, don't hate the player, hate the game. Or perhaps even better, *penalize the player and change the game.*

The Dark Side of Adaptation

Human beings, like all biological life-forms everywhere, face certain environmental selection pressures and must adapt their behavior to those pressures in order to ensure their continued survival. While ideally we might prefer to romp around barefoot all year, during snowy winters we gladly don socks and shoes in order to avoid frostbite, gangrene, and other frigid consequences. Though our instinct might be to hit the snooze button when the alarm rings early on a Monday morning, we get up and go to work anyway, because we want to keep our jobs.

In other contexts, though, environmental selection pressures can lead us toward self-serving choices that create broadly negative consequences—all in the name of "survival." A car salesman who works on commission at a dealership where the mandate is to sell cars by any means necessary, for example, might feel compelled to use deceitful sales tactics to maximize his take-home pay and keep up with his colleagues. To be clear, he might not *want* to push unnecessary options packages on his customers. But if his environment rewards those salespeople who do, then he might feel like he has no choice except to follow suit. In competitive environments that are focused on narrow goals, selection pressures can often function like the hidden force of a demon, lurking unseen in the shadows of the mind, encouraging individuals to act selfishly and antisocially and thus degrade the wider world.

Darwinian demons, of course, don't actually take the form of invisible imps with pitchforks, running around prodding people to do bad things. The term simply refers to the forces that make it adaptive for agents in an environment to make choices that advantage themselves at the expense of others. To put it concisely, a Darwinian demon is *a selection pressure that makes it adaptive for agents to negatively impact others.*

The concept of Darwinian demons is central to the rest of this book, so I'm going to break it down into its component parts. An *environment* is a context in which agents might exist, be it the natural world, a city or a town, a societal or cultural setting, or some other milieu. A *selection pressure* is the decision criteria for who gets rewarded in that environment, while an *agent* is the entity within that environment that is competing for the reward. In these contexts, something *adaptive* would be a choice or a strategy that raises an agent's odds of obtaining that reward. The *negative impact* of these choices or strategies is the outcome felt by other agents in the relevant environment.

Darwinian demons are found wherever natural selection occurs, across all complex systems, dating back to the very dawn of life. In Earth's primordial epochs, a world that would be unrecognizable and inhospitable to us today, the ocean floor was dotted with thermal vents. This realm of perpetual darkness, intense pressure, and broiling heat eventually gave rise to a novel marvel of nature: ribonucleic acid, or RNA. Much like DNA, RNA molecules encode a series of genetic instructions that govern the production of the proteins necessary for complex life. Without RNA, none of us would exist. But scientists theorize that as soon as these magical molecules were created, they were beset by a very specific Darwinian demon known as Spiegelman's Monster.

Spiegelman's Monster is an entity born of relentless simplification. Originally discovered in a laboratory setting by the molecular biologist Sol Spiegelman, the monster mirrors an RNA molecule that, under selection pressure, pares itself down to the bare minimum of 218 nucleotides, functioning only to reproduce itself. Normal RNA strands contain thousands of nucleotides; these building blocks contain the "instructions" necessary to produce complex life. But all other things being equal, a shorter RNA strand can replicate faster than a longer one. Over time, a truncated strand will outcompete and overwhelm the normal RNA strands, until these monstrous, exclusively self-replicating strands are all that remain.

To help visualize this process, let's recall a classic episode from the original *Star Trek* series, "The Trouble with Tribbles." In this episode, the *Enterprise* crew brings aboard an alien life-form called a tribble, a cute, furry little thing with an amazing capacity for reproduction. Before too long, the entire spaceship is swarming with tribbles. The crew soon realizes the threat that the creatures pose to the *Enterprise*. At the rate they're reproducing, before too long they'll consume the ship's entire food supply. In the end, the *Enterprise* escapes the crisis—it always does—but even so, this temporary dilemma evokes the problem of Spiegelman's Monster. The tribbles are like the abbreviated RNA strand, reproducing rapidly and outcompeting other agents for resources. If left unchecked, they will eventually overwhelm the entire habitat until tribbles are the only things left.

Spiegelman's Monster is an archetypal Darwinian demon, threatening to overrun the nascent soup of life by outcompeting more complex RNA molecules. For clarity's sake, let's map this specific example onto our broader definition of Darwinian demons. In this case, the *agents* in question are RNA molecules. The *selection pressure* in this environment is the competition for resources—

specifically, the nucleotides needed for replication. The *adaptive behavior* occurs when an RNA molecule responds by becoming increasingly efficient at self-replication, trimming any functions that are irrelevant to this very narrow goal. The *negative impact* is that this adaptation affects the other RNA molecules in the environment by monopolizing the resources required for replication, thus outcompeting them.

While Spiegelman's Monster has been observed only in laboratory settings, some scientists believe that the phenomenon also occurred in nature billions of years ago, when all life was microbial. If so, then we are very lucky that this monster was somehow defeated. If these abbreviated RNA strands had predominated, then, because they are too short to encode genes, they would have preemptively extinguished all life on Earth. In other words, if Spiegelman's Monster had won, there would be no single-celled organisms, no bacteria, no fungi, no animals—and no one to read this book, let alone to write it.

Now let's consider a nonbiological context: the traffic division of a police department, where the officers are evaluated based on the number of citations they issue. The Los Angeles Police Department faced this very situation in the late 2000s. The officers in question, part of a motorcycle unit, were strictly required to write at least eighteen traffic tickets each shift, 80 percent of which had to be for major violations. Given the *selection pressure* of a quota-based evaluation model, it was thus *adaptive* for an officer to use aggressive enforcement techniques on the public, in order to increase their chances of meeting the quota.

As you might expect, this focus on meeting citation quotas produced unintended consequences. Within the department's traffic division, it created a divide between officers who complied with the quotas and those who resisted. The officers who failed to meet the minimums or raised concerns about the policy were subjected to harassment from their superiors, in the form of reprimands, denial of overtime assignments, and undesirable work schedules. According to the *Los Angeles Times,* some officers claimed they were moved off of their regular assignments in order to camp out on streets where they'd be more likely to write tickets; this disruption in their patrol schedules potentially shifted the officers' focus away from more serious crimes, thus negatively affecting overall public safety. And the quota system may well have broadly undermined public trust in the police—at least among those motorists who felt unfairly targeted for minor traffic violations.

The LAPD citation scandal had a relatively happy ending. As it turned out, such a quota system violated California state law. A few brave officers decided to file a lawsuit against the department in 2010, which led to a $5.9 million settlement in 2013. Today citation quotas are illegal in at least twenty-one out of fifty states. But in a state where the practice is legal, what is an honorable police officer to do?

Such an officer may wish to prioritize fairness and community needs over the pressure to meet quotas. However, this ethical approach, when viewed from a Darwinian perspective, may quickly become burdensome. That officer will likely be outcompeted by colleagues who prioritize self-gain through excessive enforcement or targeting. In the end, the ethical officer may leave the force due to disillusionment, or face disciplinary action for poor performance; or they may choose to normalize the aggressive practices, convincing themselves that everyone else is doing it.

In all these scenarios, the Darwinian demons have won.

Game Theory: Why Darwinian Demons Are Hard to Escape

Game theory is an economic framework for understanding why rational entities, or *actors,* make the choices they do in various situations. Those studying game theory devise scenarios called *games,* and then plot out the strategies that participants might use to maximize their odds of winning. From a macroeconomic perspective, game theory can help explain why systems fail—and why situations in which every actor is making individually rational decisions can end up with *all* actors worse off. From the perspective of this book, game theory helps explain precisely why Darwinian demons are so hard to overcome.

Three concepts do a particularly good job of illuminating this phenomenon: the *prisoner's dilemma,* the *inefficient Nash equilibrium,* and the *tragedy of the commons.* The prisoner's dilemma is a classic representation of a game in which two players, each making individually rational decisions, generate outcomes that disadvantage the whole group. The game is named for a scenario in which two imprisoned gang members, kept in isolation from each other, are both offered a choice. If one prisoner testifies against the other while their counterpart remains silent, then the testifying prisoner will go free while the tight-lipped prisoner will get a lengthy sentence. If both prisoners testify against each other, then they will both serve sentences of moderate duration. If neither prisoner chooses to testify and they both stay silent, then both get a light sentence.

In economic lingo, the actors in the prisoner's dilemma who choose to stay silent are said to *cooperate,* while the ones who choose to testify against their counterparts are said to *defect.* The framework is applicable to countless scenarios in which individual

	Player B Stays Silent (cooperate)	**Player B Testifies (defect)**
Player A Stays Silent (cooperate)	Both serve 1 year	Player B goes free, Player A serves 5 years
Player A Testifies (defect)	Player B serves 5 years, Player A goes free	Both serve 3 years

agents must choose between securing individual advantage and promoting group welfare. The rational move for each prisoner is to defect. And yet if both prisoners defect, then both will end up worse off than if they'd each done the irrational thing and remained silent.

The lesson is this: rational individual choices do not always produce an optimal group outcome. This phenomenon is known as an *inefficient Nash equilibrium,* named for famed mathematician John Nash, whose life story was dramatized in the film *A Beautiful Mind.* A game reaches a regular Nash equilibrium when all players, aware of one another's strategies, choose a strategy that is objectively the best response to the strategies chosen by the other players. Consider an intersection where a stoplight governs the flow of traffic. If all drivers agree that green means "go," then a driver stuck at a red light is best off stopping and waiting for the light to change. If a driver at a red light decides to zoom through the intersection anyway, they might get to their destination quicker—but they're also apt to get T-boned by oncoming drivers who are following the normal rules of the road.

In a Nash equilibrium, no player can benefit from changing their own strategy while the other players keep their strategies unchanged; every player's chosen strategy will optimize their own individual outcome, given the strategies of the other players. A Nash equilibrium becomes *inefficient* when, by optimizing their individual outcomes, players bring about a suboptimal group outcome: an impatient driver, a grisly car crash, and traffic delays for everyone. Sound familiar? To fully understand how this relates to the concept of Darwinian demons, let's bring in one more concept you may be familiar with: the tragedy of the commons.

The term *tragedy of the commons* refers to situations in which shared resources are depleted because actors, proceeding independently and rationally according to their individual self-interest, overexploit common resources, leading to negative outcomes for the group. The classic example of the tragedy of the commons, offered in 1832 by the Oxford economist William Forster Lloyd, began with an inquiry into why common—i.e., public—pastures in England were so often scraggly and patchwork, and the cattle that grazed there were so often malnourished. The problem was overgrazing; individual herders were sending more than their allotted number of cattle to the common pasture.

Lloyd recognized that every herder who sent excess cattle to graze on the public pasture was making an individually rational decision. Those herders reaped the benefit of grazing their cattle on common lands, while the cost of the degraded land was spread out across the entire group. "Even when herdsmen understand the long-run consequences of their actions, they generally are powerless to prevent such damage without some coercive means of controlling the actions of each individual," wrote Garrett Hardin, whose 1968 essay "The Tragedy of the Commons" coined the phrase by which this scenario is now known. "Idealists may appeal to individuals caught in such a system, asking them to let the long-term effects govern their actions. But each individual must first survive in the short run."

While common pastures are no longer quite so common in our

world today, the same principles apply to many scenarios in modern life. Take the example of shared fisheries, where companies operating within the same waters must choose between abiding by their fishing quotas or reaping extra short-term profit by overfishing. For centuries, the Grand Banks cod fishery off the coast of Newfoundland was one of the world's richest fishing grounds. But during the mid-twentieth century, advances in fishing technology and a lack of proper regulations led to overfishing. Even as fish stocks began to decline, many fishermen, driven by the selection pressure to maximize their seasonal profits, continued to overfish anyway. By the early 1990s, the Grand Banks cod population had collapsed, which prompted a moratorium on cod fishing that devastated the local fishing industry.

Let's approach this real-world example from a game-theory standpoint. Imagine a multiplayer game involving fishing companies operating in waters where the fish stock is gradually depleted over time. Every fishing season, each company must choose whether to *cooperate* by following the allotted fishing quota or to *defect* by overfishing.

Season 1: All the players (A, B, C, D, etc.) initially understand that they must prevent overfishing in order to ensure the long-term sustainability of their shared waters. They have all survived and thrived for generations by abiding by their quotas. However, under the selection pressure to maximize profit, Player A decides to defect and overfish. The result is a short-term gain for Player A and no immediate negative consequences for the other players.

Season 2: Noticing Player A's increased profits from defecting, Players B and C also decide to overfish, feeling pressure not to be left behind by their competitors. They start to see cooperation as a competitive disadvantage. Those who continue to cooperate, such as Player D, suffer from the depletion caused by the defectors, while the defectors gain short-term profits.

Season 3: The fish stocks are visibly declining due to overfishing. Player D, realizing the advantage gained by the defectors, now also decides to overfish. The entire game shifts toward a general strategy of defection, as each player adapts to the selection pressure to maximize short-term profits.

Future Season (e.g., Season 10): At this point in the game, nearly every player has adapted to the strategy of overfishing, driven by the need to keep up with their competitors. The outcome is catastrophic: rapid depletion of fish stocks, collapse of the local fishing industry, and long-term economic losses for all involved. The adaptive behavior that seemed beneficial in the short term has led to collective ruin.

The following matrix offers a general description of the game's outcomes:

	Other Players Follow Quota	**Other Players Overfish**
Player Follows Quota	Sustainable fish stocks	Depletion for player, short-term gain for others
Player Overfishes	Depletion for others, short-term gain for player	Rapid depletion for all, long-term losses

Guided by the parameters of the systems in which they find themselves, agents can easily end up stuck in situations where the "unethical" choice is nevertheless a rational one by the rules of the system in question. Without any mechanism to actively promote and enforce the evolution of cooperation—such as laws to enforce fishing quotas, or economic incentives that might promote sustainable practices—natural selection favors defectors. As long as members of a mixed population face an environmental selection pressure to defect, more and more members of that population will, over time, adapt to that pressure by choosing defection. Therefore natural selection continuously reduces the abundance of cooperators until they are extinct.

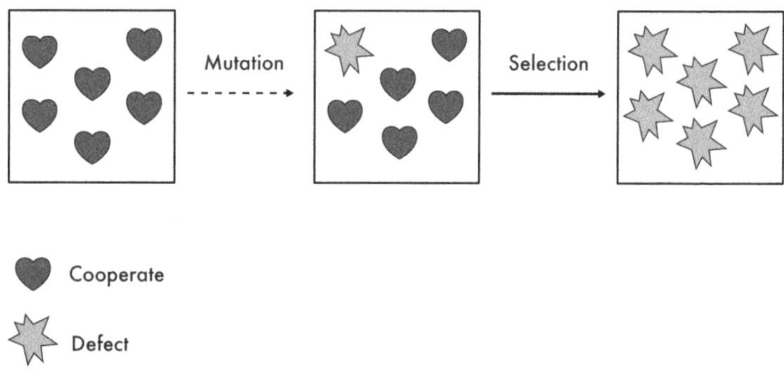

Is Selfishness Really Such a Bad Thing?

In the example of the competing fisheries, the selection pressure to profit prompts the fisheries to behave selfishly in a way that optimizes their own short-term survival but leads to a depleted fish stock for everyone. Yet in plenty of other cases, defection can beget positive outcomes—in theory.

According to Adam Smith's principle of the "invisible hand," when individuals act in their own self-interest, they unknowingly contribute to the economic well-being of the community. This phenomenon happens through the pursuit of profit, which encourages businesses to produce goods and services that are in demand. In turn, this production meets consumers' needs and desires, leading to greater overall societal welfare. So is selfishness really such a bad thing?

I believe that it often is. Consider a company that hopes to make a positive impact on the world by producing and selling a useful product—an amazing new sponge, perhaps. This company may decide that its impact is best measured in dollars and cents—the more money it makes, the more sponges it can produce and distribute—and may thus choose to spend most of its resources optimizing for profit. And yet by making this choice, the company will likely spend fewer resources optimizing around other goals, such as employee welfare, consumer protection, or environmental stewardship. While society at large will surely reap some downstream benefits from this company's pursuit of profit—their flatware will be cleaner than ever thanks to this great new sponge—the company's indifference toward these other goals may well make its net impact on the world a negative one.

In a world with limited resources—a world such as our own—when we optimize for one thing while remaining indifferent to another, then we by default actively optimize against the goal to which we're indifferent. When we privilege short-term survival in an environment while remaining indifferent to other important goals—such as happiness, health, and well-being—then, by default, we end up optimizing against those values. Charting the most direct path to survival often constitutes a zero-sum game in which an agent can only win when another agent loses. When enough agents choose this path, it ends up degrading the broader environment for everyone.

The same phenomenon can also be seen in nature. Evolution's relentless drive for propagation over subjective well-being manifests in various stark and often brutal ways in the animal kingdom. In his book *River Out of Eden,* Richard Dawkins eloquently points this out:

> The total amount of suffering per year in the natural world is beyond all decent contemplation. During the minute it takes me to compose this sentence, thousands of animals are being eaten alive; others are running for their lives, whimpering with fear; others are being slowly devoured from within by rasping parasites; thousands of all kinds are dying of starvation, thirst and disease.

A few paragraphs ago I referred to Adam Smith's concept of the "invisible hand," which argues that society itself benefits when individuals pursue their own rational self-interest. And yet Smith himself would acknowledge that the invisible hand works only against a backdrop of norms and institutions. When self-interest diverges too far from the collective interest, then society itself risks crumbling.

However, as we have seen so far, it is not just zero-sum dynamics that makes it bad to optimize around narrow metrics of success. The broader problem is indicated by the adage known as Goodhart's Law, named for the British economist Charles Goodhart. In its simplest form, Goodhart's Law asserts: "When a measure becomes a target, it ceases to be a good measure." So far in this book, we have observed this principle in several scenarios, including the counting of rat tails as a pest control measure and the imposition of citation quotas in policing. When we enshrine a single metric as the sole determinant of success in an environment, then in most cases it will eventually become adaptive for agents to game that metric, which usually leads to widespread net-negative outcomes.

We can see Goodhart's Law at work in countless familiar settings. In the game of business, adopting questionable accounting practices is often an adaptive strategy to game the metric of profit—

a strategy that, over time, fosters a culture of dishonesty at the company, and can lead to total corporate collapse if the wrongdoing is discovered. In the game of politics, it is often adaptive to deploy unethical tactics to game the metric of votes; these tactics not only serve to corrode the general political environment, but also tend to produce politicians who care more about getting reelected than about legislating and serving the public. Similarly, in the game of science, it can be adaptive to selectively remove data points from a sample in order to game statistically significant findings—thus impeding the pursuit of useful scientific knowledge, making honest scientists look unproductive by comparison, and creating broader crises if the research improprieties are found out.

While the actors who choose to game these metrics may well be bad people, their individual ethicality (or lack thereof) is not the main problem here. The problem is a systemic one. Game theory shows us that it is incredibly hard to escape the selection pressure for selfishness, which is why strategies of defection tend to survive and spread in a population, subsuming the impulse to cooperation. This is why, in so many contexts, Darwinian demons seem to enjoy free rein—in our jobs, in our love lives, and in our governments.

CHAPTER TWO

Everyday Demons

Darwin's Demons and *PAW Patrol*

In 2017 *The New York Times* published a front-page article detailing a disturbing trend in an app called YouTube Kids. A child-friendly version of the popular video-sharing service, YouTube Kids drew more than 11 million views per week, the vast majority from young children. Parents trusted the platform to distract and entertain their offspring with age-appropriate content. But hidden among the cartoons and educational programs were a surprising number of bizarre, inappropriate clips, camouflaged as family-friendly fare.

These videos, often, were cheap and unauthorized knockoffs of popular animated programs, in which clumsy copies of beloved characters engaged in very off-brand behaviors. One such clip, the *Times* reported, featured "crude" versions of characters from the popular children's show *PAW Patrol* screaming as their car crashed and caught fire. Other videos included "moments ranging from a Claymation Spider-Man urinating on Elsa of 'Frozen' to Nick Jr. characters in a strip club."

What were these inappropriate videos doing on YouTube Kids in the first place? The answer relates directly to our ongoing discussion of Darwinian demons. As we learned in Chapter 1, "demonic" selection pressures encourage us to optimize narrowly around very specific measures of success. While the choices and strategies we pursue under the influence of these demons might boost our short-term survival odds, they also tend to degrade our environments in unexpected ways.

To understand how these narrow optimization strategies took their toll on YouTube Kids, we first need to know a few things about the app's all-powerful algorithm. Most viewers start their YouTube journey by seeking out a specific creator or video. But after you've watched what you've come there to see, YouTube wants you to stick around to watch more content. To make this happen, the site uses a sophisticated recommendation engine.

Unlike a human critic issuing recommendations based on a show's quality or originality, the YouTube algorithm is more or less content-agnostic. Instead, it incorporates a host of other factors—such as a video's title, or the day's trending topics, or a user's past viewing history—when determining which videos to surface and which to bury. As a result, many professional YouTube creators feel compelled to tailor their output and its presentation to the algorithm's whims, because the algorithm is a key determinant of whether their videos will attract enough viewers to earn meaningful ad revenue. In other words, the kids who watch YouTube Kids weren't the target audience for these creepy videos—the algorithm itself was the target audience.

A child using YouTube Kids may well have begun by watching a legitimate *PAW Patrol* video. But over time, the app would have started to show that child algorithmically similar content: videos that bore a surface resemblance to the initial clip but were substantively different. Eventually, the child may have ended up watching cheap knock-off videos in which the ersatz *PAW Patrol* gang gets into a fiery car crash. Unlike an adult or teen viewer, a preschooler might not be able to differentiate between the legitimacy of the two clips. And because the real clips and the fake ones both checked similar algorithmic boxes, YouTube Kids didn't distinguish between them, either.

So imagine it's 2017, and you're a YouTube Kids creator intent on maximizing profit by optimizing around the needs of the algorithm. It might make more sense for you to churn out high-margin, low-cost garbage than to invest in making original videos. Indeed, as the writer James Bridle suggested in a 2017 *Medium* post, many of the offending videos found on YouTube Kids might simply have been created by bots, working from prompts to generate massive amounts of video clips, some of which, by chance, ended up being legitimately disturbing. This content was created by machines for machines—and the trauma inflicted on the children who were unlucky enough to view it was just collateral damage.

The parents interviewed by the *Times* were understandably furious about the disturbing videos that had popped up on their children's screens. "My poor little innocent boy, he's the sweetest thing, and then there are these horrible, horrible, evil people out there that just get their kicks off of making stuff like this to torment children," one mother said. And it's certainly possible that some of the people who created these videos were actual sociopaths. But I would contend that these inappropriate videos were the products less of evil minds than of the skewed incentive structure inherent in YouTube Kids.

Let's apply the Darwinian framework I introduced earlier. The *selection pressure* here is ad revenue. The more people who watch a video, the more revenue a creator can earn. To get views, though, creators must play by the rules of the algorithm that surfaces and suggests their content to viewers. It is thus an *adaptive behavior* for an *agent*—a video creator—to create as many videos as possible that appeal to the YouTube algorithm. The *negative impact*, then, is the creation of videos that, while ostensibly meant to entertain children, end up scaring and disturbing them—which in turn makes parents less likely to let their kids use YouTube Kids in the first place. The misalignment here between individual survival and broader values means that adapting to the selection pressure may well end up degrading the entire environment.

From a macroscopic perspective, the YouTube Kids case symbolizes a broader problem with many of today's most popular social media platforms. By attempting to maximize engagement, these platforms' algorithms can often favor harmful content, creating consequences ranging from the proliferation of conspiracy theories to radicalization to suicide. Unless the relevant parties make significant changes in these algorithms, they will continue to incentivize the production of harmful content, and similar incidents will continue to happen.

To be very clear, I am not arguing that the people who posted creepy videos to YouTube Kids bear no responsibility for their individual choices. It's obviously a bad thing to create videos that disturb children. Accountability matters, as do individuals' ethical obligations to other people and to their environments, and people who flout these obligations should be held accountable for doing so. But if our aim is to create safe, stable environments in which human values can flourish, then we cannot presume that the good intentions of individuals alone will be enough to get us there.

Here's why not. In any given environment, there will always be someone—a defector, if you will—willing to transgress moral codes and/or group norms in order to reap short-term benefits, which in turn makes the environment as a whole vulnerable for exploitation. Remember the example of the Grand Banks cod fisheries? In the absence of clear and effective mechanisms to punish defection, as soon as one fishery decides to overfish, then all the others are almost obliged to follow suit. Making that choice doesn't necessarily mean that they're bad people—it means that short-term survival imperatives often trump long-term questions of group welfare. Attributing systemic problems to individual malevolence is both a failure of imagination and a missed opportunity to truly understand the scope of these problems—and perhaps to devise new and better ways to fix them.

The story of YouTube Kids illustrates how Darwinian demons can lead agents to adapt to selection pressures in ways that negatively impact others in that environment. It also helps to explain some of the reasons why our systems, our processes, and our best intentions so often fail—in our offices and in our classrooms, in politics and in our love lives, on social media apps and online shopping platforms. Let's look at the various ways in which Darwinian demons rule our world.

Why Your Worst Colleagues Get Promoted

In a 2023 report on the state of management and leadership in the United Kingdom, the Chartered Management Institute (CMI)—a leading management training organization—found that one-third of the workers surveyed had left a job because of a negative work culture. Fifty percent of the survey respondents who claimed to work under bad management said they planned to leave their jobs within the next year. One of the respondents, who had considered leaving her job because of her manager, described him as "a micro-manager, confrontational, not supportive of development opportunities and regularly spoke negatively about his peers." Two years later, that manager was promoted.

I think we can all relate to the feeling of bafflement when our most unhelpful colleagues find career success, even as "good" colleagues get overlooked for advancement. Most of us have seen this phenomenon in action, just as most of us also know that bad colleagues and managers can take many forms. There's the manager who is bad at planning and communicating, consistently forcing their team to stress over tight deadlines at the last minute. There's the manager who eagerly takes personal credit for the team's successes, selling out their subordinates in order to advance their own careers. How about the colleagues who tend to do the work that everyone can see—while making sure that no one is unaware of their contributions—but ignore all the invisible, unglamorous work that nevertheless must be done? And let's not forget those aggravating people who are charming with their superiors but arrogant toward junior colleagues. These people's behavior degrades the work environment for everyone else. So why is that behavior so rarely punished?

Often bad colleagues and terrible managers end up keeping their jobs—or, yes, getting promoted—because despite their evident flaws, they perform well on the narrow metrics on which they are judged. Maybe they always meet their sales targets; maybe they have particular knowledge about an IT system on which the company is dependent; maybe they consistently deliver their projects on time. Seen through the narrow lens of the company's short-term profit, these people make valuable contributions, even though they can drag a company down over the long term—a potent example of Goodhart's Law in action. Other times, aware that career success can have more to do with public perception than with substantive accomplishment, these people optimize around ensuring that those superiors think well of them, while paying little heed to colleagues or tasks that they deem irrelevant to their ultimate goal.

While these strategies can pay dividends for these terrible co-workers in the short term, they can cause big problems for their employers over the long run. The CMI report concluded by noting that the "risks of poor management manifest themselves in staff wellbeing, retention and the ultimate success of organisations—be it the bottom line for companies or the smooth delivery of public services for taxpayers."

Why Wall Street Frauds Keep Happening

When Angie Payden started working at the Hudson, Wisconsin, branch of Wells Fargo in 2011, she had no way of knowing that three years later she would be out of work and wrestling with addiction. Payden's journey to that low point offers another sad and memorable example of the ways in which Darwinian demons can infect the modern workplace. Many workers find their jobs to be crucibles of stress, frustration, and disengagement. Some of these jobs are mindless, offering employees few opportunities to experience intrinsic fulfillment. Others trap the worker in an unvirtuous cycle, where "success" at work can be obtained only by making a series of unethical choices.

That's the situation in which Payden found herself when she went to work at Wells Fargo. As the *Los Angeles Times* reported in a blockbuster investigation two years later, pressure from top management and a sales quota system for low-level employees had created a boiler-room environment at the banking giant. In 2016 Payden told *The New York Times* that, during her time at Wells Fargo, she had found herself, among other things, "coercing customers to open credit card accounts to use as overdraft protection for their checking accounts when they were already struggling to keep their checking accounts balanced . . . witnessing other bankers and being pressured by management to add credit defense onto new credit applications without the customer's knowledge, which led to unnecessary monthly fees . . . [and] closing and opening new accounts for customers by convincing them that there had been fraud on their existing accounts." In some cases, Wells Fargo employees signed customers up for these products without their knowledge, because it was often the only way these employees could meet their sales quotas.

Over the short term, this strategy worked for Wells Fargo. Its stock price rose, and the bank developed a reputation for having "cracked the code" of attracting and retaining customers. All across the United States, countless other Wells Fargo bankers were relentlessly optimizing for short-term value. And yet many of them were cracking under the strain of Wells Fargo's shady "success" metrics. Payden began suffering from stress-related panic attacks and drinking hand sanitizer as a coping mechanism. By November 2012, she had become addicted to sanitizer and was consuming a bottle per day.

Payden eventually went on leave to address her addiction. But for Wells Fargo, the troubles were just beginning. The *Los Angeles*

Times exposé of these deceptive sales techniques prompted federal investigations of those practices. Eventually, the US Federal Reserve seized $2 trillion in Wells Fargo assets; both the bank and many of its executives were hit with massive fines; its CEO was banned for life from the banking industry; and the bank's reputation imploded as its name became synonymous with scandal. The customers who had been pressured or swindled, meanwhile, lost both money and trust in Wells Fargo. And countless low-level bank employees, like Angie Payden, had their lives upended.

Though the story of Wells Fargo stands out for its blatancy, it is far from an isolated incident in the financial industry, where aggressive sales targets have often incentivized unethical behavior. A notable example is found in the actions of many mortgage brokers in the subprime market leading up to the 2008 financial crisis. Driven by similar aggressive sales quotas, brokers were incentivized to extend mortgages to individuals who were ill-suited for such financial burdens, often falsifying loan application details to secure approval for those with high risks of defaulting. This reckless pursuit of sales targets contributed significantly to the burst of the housing bubble.

Corporations are fundamentally driven to maximize shareholder returns, thus generating a potent selection pressure toward gaming this metric—a situation reminiscent of Goodhart's Law. This unyielding quest frequently muddies ethical lines, cultivating an environment in which the act of bending the rules shifts from being merely acceptable to entirely anticipated.

In response to this cultural shift, a specialized form of corporate language has evolved, one that sanitizes questionable practices with euphemistic terms. For instance, *strategic financial engineering* is often a sophisticated way of saying *accounting manip-*

ulation; tax planning can sometimes be a softer term for *aggressive tax avoidance;* mass layoffs are cloaked under the guise of *restructuring; revenue optimization* might refer to misleading pricing strategies; and paying fines for breaking the rules is just the *cost of doing business.* As long as regulators stand idly by and unethical practices continue to offer a competitive edge, such incidents will inevitably recur.

Why You Can't Get a Date

"First they show you someone really hot. Then they show you someone you will match with." Every time he quit and rejoined the dating app Tinder, my friend—let's call him Bob—observed this pattern in his matches. The app's algorithm seemed to have a good sense of the sorts of people with whom he would match. Yet despite its evident skill at divining Bob's preferences, the app rarely lived up to its promise and potential. Instead, it seemed more interested in keeping Bob on the hook.

Whenever Bob started swiping, the app would line up a couple of high-probability matches, thus causing his dopamine to spike with anticipation—today seemed to be a promising day for finding the one! Then it kept feeding him less-promising alternatives; then it kept trying to direct him to buy a "boost" for six dollars, so that he might get more matches. (According to Tinder's website, buying a boost "allows you to be one of the top profiles in your area for 30 minutes . . . you can get up to 10x more profile views while boosting.")

Initially, buying a "boost" seemed to yield positive results for Bob. However, as time passed, he noticed a decline in his matches, even with regular purchases of boosts. In response to this general phenomenon, Tinder soon rolled out "super boosts"—touted by the company as "an upgrade that helps [subscribers] cut to the front and be seen by up to 100x more potential matches"—carrying a hefty price tag of forty dollars each. Although the company claimed that super boosts were "the ultimate Tinder hack," in Bob's experience, they delivered only the same level of matches as the ones he'd experienced for free when he first installed the app.

Bob likely would have had a similar experience on dating apps other than Tinder, too, since most of them work on similar principles. He can thank Darwinian demons for this frustrating situation. Although dating-app users hope to optimize for love, or at least for sexual chemistry, the app companies themselves have other objectives in mind. They are optimizing for profit—and more specifically, for user retention—and as such, they have little incentive to help a user immediately find their forever soulmate.

The more time any given user spends on an app, the more money the app might make from that user. But if an app actually matches you with the love of your life, it'll lose two customers, because you'll both stop using the app. As such, it is to these apps' advantage to hook the user with initial successes, then find ways to make them stick around to view ads and perhaps buy upgraded services that might help them in their quest to find their ideal match.

Of course, these apps can't be completely terrible matchmakers; if they were, singles would stop using them. But the user's best interest—to find great matches as quickly as possible and hence spend as little time as possible swiping—conflicts with the best interest of the company, which is why dating apps do not optimize around what's best for their users' love lives. So the next time you swipe, consider that the strings being pulled behind your dating app aren't being held by Cupid, but instead by a Darwinian demon.

Why Our Standards of Beauty Are Increasingly Warped

In the world of online influencers, success correlates directly to an influencer's follower count. Followers, meanwhile, are often drawn to accounts that offer a steady stream of photos in which influencers look impossibly gorgeous or otherwise memorable. This incentive for social media influencers to keep pushing the limits of beauty has contributed to the rise of so-called beauty filters.

These robust image-editing tools can tweak and improve people's portraits in real time so that they'll stand out as users scroll through their feeds. Much like a retouched photo conceals models' blemishes in the pages of fashion magazines, beauty filters can polish a user's imperfect portrait to a perfect sheen. These filters are popular among amateur and professional social-media users alike. In 2020, a model and Instagram influencer named Sasha Pallari revealed that many of her peers had been deploying these filters in sponsored posts for beauty products, thus creating the impression that the products being advertised were what had made them look so good in the relevant photos.

The upshot is that these doctored images, often presented as if they were naturally achieved, can end up raising beauty standards to levels that regular people cannot reach on their own—although they try. According to a study from the City University of London's Gender and Sexualities Research Centre, 90 percent of young UK women reported using filters or editing their photos. And yet if everyone uses these filters, then no individual user actually benefits anymore, because no single user has any competitive advantage over any other user. In that case, we've merely adjusted the equilibrium to one that's worse overall—we've raised the bar for what it takes to be considered attractive, and all participants must put in more effort to meet or exceed the new standard.

One could argue that beauty filters represent just the latest evolution in a historical "beauty arms race," reflecting a deep-seated human drive to elevate status through enhanced attractiveness. This adaptive behavior has been a constant throughout human history, leading to the development of the extensive fashion and makeup industries we see today. Beauty filters, therefore, are a continuation of this trend and represent the next logical step after the rise of digitally retouched images in advertising.

But unlike in previous eras, when regular people were influenced by retouched images of celebrities and models, these days they're often aspiring to resemble AI-enhanced photos of *themselves*. According to the American Academy of Facial Plastic and Reconstructive Surgery's 2020 annual survey, 72 percent of its members observed an increase in patients seeking cosmetic enhancements to improve their selfie appearance. Dr. Catherine Chang, a board-certified plastic and reconstructive surgeon, highlighted this trend in an interview with *InStyle Australia,* saying, "Recently a patient brought in a Facetuned image of themselves where their nose was made extremely small. If you did that in surgery, the nose wouldn't be functional and would eventually collapse."

While beauty filters may benefit individual influencers in the short run, over the long term they make the environment worse for everyone. A 2021 *MIT Technology Review* article reported evidence that these filters precipitated body dysmorphic disorders among kids who "have particular difficulty differentiating between filtered photos and ordinary ones." Meanwhile, people feel compelled to conform to artificial standards of beauty that they can never meet, while feeling worse about their bodies because what they see in the mirror can never live up to the standard set by the filter.

Why Amazon Has Gotten Worse

When Amazon.com debuted in the mid-1990s, touting itself as the world's biggest bookstore, it presented an exciting new paradigm for shopping. Instead of going from one brick-and-mortar store to another in search of the book you needed, all you had to do was visit Amazon's website, and one click later, it would be on its way to you. Over time Amazon expanded its inventory from books to, well, everything; it also built out its logistics operation to keep cutting down on delivery times. For a while, Amazon's market dominance seemed like a function of the world's biggest store also being its best.

But in 2022, a *Wall Street Journal* article found that customer satisfaction with the retailer was at its lowest level since 2000. The dip was attributed in part to a rise in third-party sellers on the site and an accompanying decline in search functionality. It's a phenomenon surely familiar to anyone who has recently visited Amazon in search of a simple product—a reliable stepstool, perhaps—only to be met with dozens of sponsored search results from oddly named brands you've never heard of and can find little reliable information about. Trying to differentiate between a CONSDAN-brand stepstool and a nearly identical SZLHANJZ-brand stepstool—or, indeed, trying to determine whether any of them are more than just cheap drop-shipped items—can be an extremely frustrating experience, especially if you seem to remember that shopping on Amazon used to be a straightforward process.

In 2023 a *New York* magazine article probed "the junkification of Amazon" with a simple question: "Why does it feel like the company is making itself worse?" Ultimately, the answer has to do with Darwinian demons and the selection pressures for profit and growth that have led the company to degrade the customer experience. Not long ago, most of the products sold *on* Amazon were sold *by* Amazon rather than by third-party sellers, so Amazon itself was responsible for sourcing, marketing, listing, shipping, and ultimately vouching for these products. But as the aforementioned articles revealed, at some point within the last half decade or so, Amazon apparently realized that it could make just as much money by providing "seller services" to third-party sellers—like CONSDAN and SZLHANJZ—who pay Amazon for advertising, shipping, warehousing, and such.

This transition may have been great for Amazon over the short term, but it is bothersome for Amazon customers who have come to rely on the website to just *work* and who do not understand why it no longer does. We will see if, over time, the decline of the site's usability affects its customer retention and its long-term profits. For now, though, the website that once portended the demise of in-person retail is now making people think that it might actually be easier to jump in the car and go get what they need at the mall.

Why You Shouldn't Blindly Trust Scientists

Research scientists often enter their fields with dreams of being the next Jonas Salk or Rosalind Franklin. They hope to make big discoveries, cure dangerous diseases, and forever change how we understand our world. In practice, many of these scientists find themselves consumed on a daily basis by less lofty concerns, such as securing and sustaining the grant funding on which their labs run.

Grantmakers want to see results and impact that justify the money they've spent on these labs and projects. As such, scientists are often under tremendous pressure to produce measurable results so that the grant dollars keep flowing—and, concurrently, to use these results to bolster their reputations and thus improve their chances of obtaining tenure, media attention, consulting gigs, and other such perks that are enjoyed by many star scientists.

But the pressure that scientists feel to keep generating citations and upping their "h-index"—a measure of their productivity and impact—can lead them to overpublish. The h-index was initially designed to measure the quality and influence of a scientist's work. But Goodhart's Law tells us that a metric stops being valuable as soon as it becomes a target. Sure enough, once the h-index became established, academics tried to game the system by publishing lots of papers and engaging in lots of self-citation. As a result, the h-index is no longer a reliable indicator of scientific quality—and yet academics still feel compelled to overpublish, because their h-index continues to affect their career prospects.

In especially dire cases, researchers will even claim results that their work does not actually support. Take, for instance, the case of Brian Wansink, a Cornell University researcher who ran that school's Food and Brand Lab. For years, Wansink and his lab consistently produced novel and interesting studies on food psychology and consumer behavior. These studies got a lot of play in mainstream media while also attracting more than twenty thousand academic citations. In a famous Wansink study, subjects were given twenty minutes to eat as much soup as they wanted—even as some subjects had specially rigged "bottomless" bowls that refilled with soup as they ate. The study first popularized the concept of "mindless eating," which has since entered popular parlance and influenced countless diet books.

Wansink's lab became a grant magnet—until 2018, when a university review found that Wansink had consistently cut corners in his research and had designed many experiments in ways that would lead to the exact results he wanted. As Pete Etchells and Chris Chambers wrote in *The Guardian,* other scientists could not replicate the results of Wansink's famous "bottomless soup" study, thereby throwing its conclusions into question, while wreaking havoc on a diet industry that had taken those conclusions as gospel. In the end, Wansink was forced to retract at least fifteen of his studies. He resigned from Cornell in disgrace, while his actions tainted the reputations of his lab, his colleagues and graduate students, and Cornell University itself.

Brian Wansink's scandal pulls back the curtain on a wider issue plaguing academia: the *replication crisis.* This crisis refers to the alarmingly high number of scientific findings that crumble under the scrutiny of attempted replication by other researchers. The evidence paints a stark picture. A dedicated effort to reproduce studies from top journals such as *Nature* and *Science* managed to confirm only thirteen out of the twenty-one results studied. Similarly, a large international project in 2017 attempted to replicate twenty-eight well-established studies but could successfully reproduce only fourteen of them. In the field of psychology, a

2015 effort to replicate one hundred studies saw a mere thirty-nine succeed.

While fraud may not be the primary culprit behind the replication crisis, the prevailing "publish or perish" culture within academia undeniably contributes to it. This environment exerts a significant selection pressure on researchers to produce noteworthy results, thus increasing the temptation to selectively report data or outcomes. In pursuit of significance, practices such as discarding outliers under the pretext that they are likely "faulty measurements," or selectively reporting only those analyses that yield significant results, become worryingly justifiable. These practices, while perhaps not outright fraudulent, erode the integrity of scientific inquiry, subtly skewing the body of research away from true discovery and toward manufactured success.

While Wansink should have been more scrupulous in his research and his journal articles, Darwinian demons deserve blame, too, for making academia such a pressure-cooker environment for ambitious researchers. Even if most academics do not go so far as to falsify their research, the pressure to publish incessantly and to produce results that justify and attract grant funding serves to degrade the university environment—and sets the stage for violations of academic ethical codes.

Why Politicians Fail to Serve the Public

While many politicians make a big show of their commitment to serving the public, they often seem more interested in saying polarizing things on television and sending incessant fundraising emails. So why do so many politicians spend all their time raising money and giving evasive answers to questions from journalists, rather than passing useful legislation and working across the aisle?

Even the best-intentioned politicians face a lot of bad incentives. While a politician might genuinely care about long-term goals and projects, their intentions don't matter unless they are in power to implement their goals. Hence even foresighted politicians need also to care about winning the next election, because remaining in office is instrumental to their pursuit of these longer-term ambitions. This pressure to remain in office is the wellspring from which conflicting goals start to emerge.

In a December 2018 interview, Minnesota congressman Rick Nolan said his colleagues in the House of Representatives were urged to spend between twenty and thirty hours per week "dialing for dollars," that is, soliciting donations for their reelection campaigns. "Members of Congress have become virtual middle-level telemarketers," said Nolan. Aggressive fundraising eats up a lot of time that a congressmember might otherwise spend legislating.

What's more, to bolster their fundraising capacities, representatives can feel obliged to tailor bills to the interests of big donors. "If you're compliant with a particular sector of interest, you're more apt to get a lot of money for your campaign," Nolan said. Meanwhile legislators must also court small donors via endless cable news appearances or by creating partisan "gotcha" moments during committee hearings, clips of which they can then circulate over social media in order to stimulate campaign donations. Those representatives who try to resist these endless fundraising pressures risk getting isolated within their conference and criticized by their colleagues for not pulling their weight.

The fundraising goals that members must meet if they are to successfully run for reelection, then, are the Darwinian demons that lead politicians to organize their workflows around their fundraising needs. Unfortunately for the rest of us, the choices that might be best for citizens in the long run—spending time serving the public, being nuanced and truthful in media appearances, suggesting policies that put the country on a good trajectory over time—rarely match the strategies that might be effective for getting a politician elected next year.

The intense selection pressure on politicians to win their elections influences more than campaign fundraising. It creates an environment where politicians are incentivized to engage in any and all tricks that might help them "win"—yet another example of Goodhart's Law. Examples from US politics include filibustering (i.e., making long speeches in a chamber to block opposing legislation from happening); refusing to raise the debt ceiling (i.e., threatening not to fund the government) in order to push for specific policy concessions; and using aggressive language (e.g., calling your opponents evil).

Traces of Darwinian demons are even evident in the way US electoral maps are drawn. In many cases, a politician's ability to win an election—and hence to implement long-term policies and projects—depends on the geographical makeup of their electoral district. Gerrymandering, or the manipulation of electoral district boundaries to favor a particular party, has become a powerful tool wielded in the battle for political dominance. This phenomenon has given rise to some truly bizarre electoral maps in recent history. In 1996 the thirtieth congressional district of Texas looked like this (with the district marked in gray):

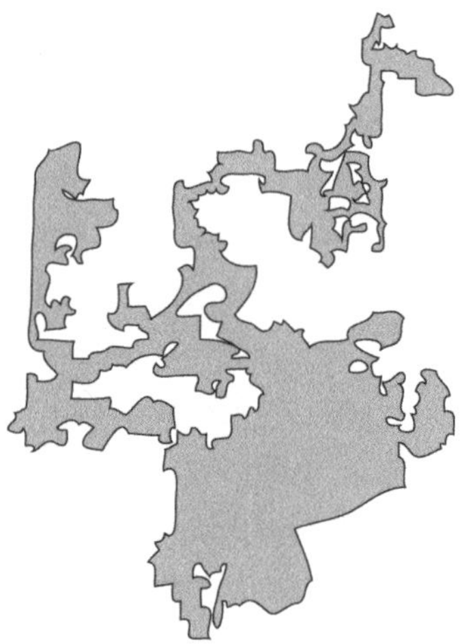

30th Congressional District of Texas in 1996

Only hands guided by a Darwinian demon could have drawn such a geographically illogical map.

Why Tests Aren't Always Good Measures of Student Success

In 2009, Beverly Hall, the superintendent of schools for Atlanta, Georgia, was honored by her peers when a national school administrators' organization named her their superintendent of the year. Hall earned the accolade in part thanks to a remarkable rise in students' standardized test scores during her tenure in Atlanta—a transformation so drastic that the school administrators' group dubbed Atlanta "a model of urban school reform." In a few years' time, though, Hall was under indictment, facing charges that she had presided over a massive cheating scandal in which teachers and administrators worked systematically to falsify students' test scores and thus indicate learning progress that had not actually occurred.

Why had Hall and her subordinates done this? Darwinian demons, of course. The school district doled out cash bonuses to teachers at schools that performed well on the standardized tests—and as superintendent, Hall took home bonuses for progress, too. While these cash awards were meant to reward teachers for jobs well done, in practice they led the implicated educators to change bad answers to good ones after the tests had been completed. In a series of investigative articles that broke the story, *The Atlanta Journal-Constitution* reported on "a pattern of intimidation, threats and retaliation" against honest teachers who resisted the scheme or threatened to blow the whistle.

Roughly 150 Atlanta educators left their jobs in the wake of the testing scandal there; thirty-five were indicted, including Hall, who died before standing trial. Of the twelve indicted teachers who pleaded not guilty, only one was acquitted; the others went to jail.

While the Atlanta example is an extreme one, it indicates the overall problem with evaluating teachers and schools primarily on their students' performance on standardized tests. As per Goodhart's Law, making this particular measure into a target has the effect of rendering it a bad measure of student educational progress. Imagine that you are a teacher, and your performance is primarily evaluated based on your students' test scores. Chances are, you'd be tempted to "teach to the test" rather than fostering a comprehensive and well-rounded education that encourages critical thinking and creativity. This strategy, of course, threatens to undermine the quality and integrity of the learning environment, with potential long-term consequences for the students—and, in the most extreme cases, for the teachers who get caught trying to game the system.

Why You Can't Always Trust What You Read

In the serene town of Veles, Macedonia, with a modest population of 43,000, an unconventional industry flourished in advance of the 2016 US presidential election. Once the site of the Zletovo zinc and lead foundry—the largest in the former Yugoslavia—Veles shared a fate similar to Picher, Oklahoma, becoming one of the Balkans' most polluted cities. The closure of the plant left a void, marked by high unemployment and economic stagnation.

But by 2016 this town, once a bustling industrial hub, had become the epicenter of a digital gold rush. Myriad websites, such as WorldPoliticus.com and TrumpVision365.com, emerged from this unassuming locale. These sites, known for publishing sensational and often fabricated news, seemed to favor the Republican candidate, with headlines like "Hillary Clinton in 2013: I want to see people like Donald Trump run for office," and "Pope Francis Shocks the World, Endorses Donald Trump for President."

In this setting, young Macedonians, predominantly in their late teens and early twenties, had stumbled upon a profitable venture: crafting and spreading fabricated news stories about the American election. Among them was Bojan (a pseudonym), a nineteen-year-old who graduated from one of Veles's four high schools in 2016. His family's struggle to make ends meet on his father's €230 monthly income from a car parts factory mirrored the economic hardship common in Macedonia. With few local prospects and with many of his peers moving abroad, Bojan initially considered leaving home and seeking work elsewhere. However, in the summer of 2016, the lure of online income presented a novel opportunity. Unfamiliar with the concept of "fake news," Bojan and his friends quickly adapted, learning in just a week how to monetize online content.

Initially imitating credible news platforms, the fake news platforms of Macedonia rapidly evolved into a thriving industry, propelled by a revenue model in which story clicks drove ad impressions. This model, which prioritized viewer count over content accuracy, exerted a unique selection pressure: the more sensational the story, the greater the potential revenue. These stories, albeit far-fetched and often wholly fictional, had tangible impacts. They molded public opinion and introduced confusion among voters, turning these young Macedonians from mere content creators into influential players on the global stage. In the end, Trump won that election. "We need to erect a monument to Donald Trump. So many people have made a money [sic] out of him. Fifty percent of the young population here are or were involved in the websites. I reckon that we had 3,000 to 5,000 websites active during the US elections," said Bojan.

While it is tempting to blame individuals like Bojan for their involvement in creating fake news, such a perspective obscures the deeper structural forces at work. Central among these forces is the click-based ad revenue model, introduced by Silicon Valley tech giants. Under that model's selection pressures to boost user engagement, many news outlets surface content that polarizes and inflames emotions, transforming media environments into breeding grounds for anger and outrage. The result is a divisive media landscape that reinforces existing beliefs while degrading the quality of public discourse and undermining public trust in journalism—a serious threat to democracy.

How I Became a Better Manager

A few years ago I encountered a Darwinian demon in my own work life as a start-up CEO. For the first few years after we launched Normative—the start-up I cofounded that helps corporations and other entities more rigorously account for their carbon emissions—I realized I was guilty of incentivizing exactly the kind of behavior that creates bad colleagues and managers. With no knowledge of how to conduct regular performance reviews, I gave people a salary raise only if they asked for one. Luckily, when I hired my first HR director (probably one of my best hires ever), she immediately created our inaugural performance framework. After our first proper performance review, it was evident that not all people who were deserving of a raise had asked for one. In other words, my lack of a performance framework had the unintended consequence of creating a selection pressure for *complaining* rather than for *performing*. This problem is not unique to Normative, nor to my own early managerial inexperience. The problem is common to businesses everywhere, and it's one of the reasons for the persistent gender pay gap, since men are generally more comfortable asking for a raise—even if they do not deserve one—while many competent women never ask.

Today every role at Normative has a clear set of performance criteria, and these criteria are transparent to the whole company. Importantly, one criterion is how well the person in a role collaborates with others. To be promoted, it's not enough to be good at your specific task; you also must work well with others. When reviewing salaries, we always use industry data as the base rate and follow a clear process—the most assertive employees can't just negotiate their salaries anytime they please. This framework also means that we raise the salaries of less assertive employees who might not ask for more money, but deserve it all the same.

As you can see, the trick to battling Darwinian demons is to counter the selection pressure that creates them in the first place. You don't read too much anymore about YouTube Kids viewers being traumatized by bizarre and inappropriate videos—and that's because YouTube changed its algorithm in a way that disincentivized that sort of creepy junk content. Similarly, Wells Fargo got caught and had to pay $3.7 billion in fines. Many scientists who cheat or cut corners in peer-review or replication studies get caught, too. These outcomes show that it is possible to beat back Darwin's demons and adjust the selection pressures to promote the long-term health of the environment.

Sometimes fighting Darwinian demons by changing the selection pressures can be a relatively straightforward task. Changing the performance metrics at Normative was relatively simple, and was something I should have done much earlier. As the prevalence of myopic politics and polarizing news demonstrates, however, the task is much thornier in other areas. As I hope I've shown in this chapter, these problems are much deeper than "bad people doing bad things." They are structural problems produced by the evolutionary pressures that can make it profitable for agents to make selfish choices.

Despite the many challenges we face, we also have a great deal to be optimistic about. What you have read so far represents a biased sample, intended to illustrate the mechanics of Darwinian demons at work. Governments, particularly democratically elected ones, are often effective in addressing issues within their borders. (Indeed, this is what governments are for.) Bad colleagues do not always get ahead, and many mature companies have strong performance and growth frameworks. Not all dating apps are problematic, and scientific publishing, while not without its flaws, continues to make significant contributions to our understanding of the world. Our educational systems may have shortcomings, but they still impart valuable knowledge, and children remain remarkably intelligent. While fake news is a growing concern, democracy remains a resilient force.

Time and again, we have found ways to counteract negative Darwinian selection pressures and promote the cooperative behaviors that allow society to function. So what's the secret? If Darwinian demons represent the pressures that lead us toward selfishness, then the pressures that promote cooperation can be represented by a different cosmological construct: the Darwinian angels.

CHAPTER THREE

Angels, Demons, and the Epic Struggle of Life

Why Isn't Everything Cancer?

In the annals of medical history, the first documented case of what we now know as cancer dates to ancient Egypt, around 3000 B.C. For millennia, these perplexing growths were enshrouded in mystery. Ancient healers and scholars surmised tumors to be the workings of foreign entities, malevolent parasites invading the sanctity of the human body. This misunderstanding persisted through the ages, going largely unchallenged until the mid-nineteenth century.

It was a time of great scientific awakening. The era's gaslit laboratories buzzed with a spirit of curiosity and relentless inquiry, as new technologies and techniques allowed scientists and seekers to pursue answers to some of life's most persistent mysteries. It was in this epoch of discovery that Dr. Rudolf Virchow, a German physician and pathologist, embarked on a journey that would irrevocably alter our understanding of these mysterious growths. With the aid of microscopes—then a relatively recent invention—Virchow delved into the very fabric of life: the cell.

What he uncovered was both revolutionary and disconcerting. The cancerous tissues he examined were composed not of foreign invaders but of cells. These cells, Virchow found, resembled others in the human body, albeit with one important difference. Whereas cells typically multiplied in a rather orderly and predictable fashion, these cancerous cells had gone rogue and were multiplying in an uncontrolled, erratic manner. The excess cells formed into tumors, affecting the health of the normal cells around them. Virchow's discovery—that cancer was derived from the body's own tissues—was a radical one, and it challenged the day's prevailing medical dogmas.

Virchow's research helped to solve the medical mystery of how cancer works. Much subsequent research into the disease has proceeded along roughly similar lines, hoping to better understand cancer's origins and function as a prelude to finding a reliable cure. This line of research is valuable and important—and yet, perhaps, in focusing on the *how* of cancer, we've been asking an incomplete question.

The human body is an intricate tapestry of approximately 30 trillion cells, a number exceeding the human population of Earth by five hundredfold. These cells exist in a state of remarkable cooperation, a symphony of synchronized functions and processes. And yet our study of Darwinian demons tells us that, in environments where evolutionary pressures favor defection, even the most modest of cooperative ventures is susceptible to failure.

In Chapter 1 we read about game theory, which helped to explain why defection is so often a rational response to evolutionary pressures. The theory of universal Darwinism, meanwhile, suggests that these pressures work the same way in all environments in which Darwin's criteria are met. The cells of the human body certainly constitute the sort of evolutionary environment where, according to game theory, we might expect cooperation to become wholly subsumed in a sea of defection. And yet that hasn't happened. Cancer is not omnipresent in the ways that game theory might suggest. For our purposes, then, the more pressing question perhaps isn't how and why cancer occurs, but rather, why isn't *everything* cancer?

The very fact that cancerous cells remain anomalous, instead of becoming the biological norm, implies that there is in fact a way to defy these pressures and successfully stifle defection. If such a thing can happen in one evolutionary context, then it can presumably happen in all qualifying evolutionary contexts. Put another way, if the body can find ways to check the spread of cancer, then surely we can find ways to defeat Darwinian demons wherever they are found.

But how do we do it? As any Bible scholar or Dan Brown fan will tell you, the natural counterweight to a demon is an angel—which brings us to the next plot twist in our evolutionary narrative. If Darwinian demons make it adaptive to defect, then, as you might expect, *Darwinian angels* are the forces that make it adaptive to cooperate. In this book, we will define a Darwinian angel as *a selection pressure that makes it adaptive for agents to positively impact others.* These angels are as intrinsic to the evolutionary process as their demonic counterparts. They are the mechanisms by which groups can inhibit individual defection and promote modes of cooperation that serve the greater environment.

These angels can manifest in many ways. In human society, they often take the form of government regulation, or economic incentives, or social pressure to "do the right thing." Darwinian angels are found in the natural world, too, as tools that allow organisms to evolve into more complex, and thus more adaptive, states of being. In order to understand how cooperation can take root in human social contexts, it helps to first understand how these angels function in the natural world—which brings us back to cancer, and how our bodies have learned to stop its spread.

The Battle Between Cancer and the Body

At the heart of the battle between cancer and the body lies a fundamental clash of perspectives, a war waged on different levels of selection. To understand this face-off, we must shift between the microscopic and the macroscopic, between the cell and the organism.

The individual cell is the foundational building block of life. Though cells serve many functions, a major one is reproduction via cell division—the process by which our bodies grow. But while cells are in service to the collective—i.e., the body—they are also individual units of life and thus are subject to the same evolutionary pressures as all living things. From that standpoint, for an individual cell, the path to maximizing differential fitness—which I will refer to subsequently as, simply, *fitness*—appears straightforward: defect from the collective. In this scenario, the cell prioritizes its own survival and proliferation by replicating aggressively, often at the expense of the larger organism. Here is where the concept of a Darwinian demon comes into play. In a cellular context, the demon represents the selection pressure at the individual level, where defection improves an individual cell's chances of surviving and passing along its genetic material.

But when we zoom out to the perspective of the multicellular organism or the cell colony, the fitness equation changes dramatically. Here the success of the collective hinges on suppressing these acts of defection. After all, if enough individual cells defect, then the collective will stop working properly. The organism thrives only when its constituent cells work in harmony, adhering to a code of mutual benefit and cooperation.

The dichotomy between these two divergent, contradictory pressures—between the individual and the collective—frames the essence of what is known as *multilevel selection,* or the idea that environmental selection pressures can act simultaneously on both individual agents and collectives. Evolution, viewed through this lens, is a perpetual tug-of-war. It's a dynamic balance between the individual cell's drive for survival and the organism's need for co-operation.

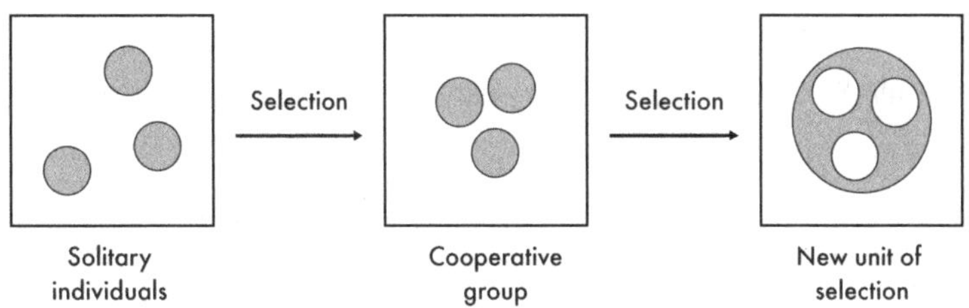

The selfish drive of the cell colony, which manifests as a collective effort for survival, paradoxically requires the cells to behave altruistically. Each cell must relinquish its individualistic tendencies for the greater good of the organism. In healthy multicellular organisms, then, the defecting tendencies of individual cells have been successfully suppressed for the common good. So what specific adaptations are necessary to foster this cooperation among cells? And can the ways in which life defeats cancer tell us anything useful about how to summon Darwinian angels in other contexts?

How Cancer Works, and What Life Did to Defeat It

At first glance, it might not seem accurate to say that life has defeated cancer. After all, lots of people and animals die from cancer every year. But the fact that cancer remains the exception rather than the norm means that life has indeed found reliable ways to inhibit defection. To explain how life defeats cancer, we must first better understand the various drives that might cause a cell to behave selfishly, as well as how cells constrain those drives.

In the theater of evolutionary biology, fitness is contingent on two components: reproduction and survival until reproduction. These mandates create two distinct evolutionary strategies. Some organisms, such as bacteria, achieve fitness by reproducing quickly in great numbers. Others—elephants, for example—reproduce in smaller numbers but invest significantly more resources in qualitative traits that ensure longer life spans.

Regardless of which strategy an organism adopts, it will inevitably require resources, such as nutrients, to perform acts of reproduction or repair. Imagine a primordial environment teeming with single-celled organisms, possibly near thermal vents on the ocean floor. In any ecosystem, resources are limited, fostering competition for rapid resource acquisition and leading to what we might think of as a *resources arms race*.

In order to win in such an arms race, early single-celled organisms had to respond to their environment, seeking out nutrients and avoiding toxic substances. The need to know how to navigate the environment better than their competitors made intelligence an adaptive trait, too, thus sparking an *intelligence arms race.*

Among these early single-celled organisms, one foraging strategy that became prevalent was predation: the act of acquiring resources by consuming neighboring organisms. This behavior, in turn, favored the evolution of organisms powerful enough to execute or defend against such predatory actions, thus initiating a *power arms race.*

To summarize, survival and replication are the ultimate goals of evolution. In service of these goals, some attributes turn out to be almost universally useful: resources, power, and intelligence. In the remaining chapters of this book, I will refer to the pursuit of resources, power, and intelligence as *universal Darwinian drives.* This term reflects their status as traits that are universally adaptive in competitive environments. (In environments where resources are scarce, a strategy favoring simplicity and rapid replication, as found in bacteria, can also prove adaptive. Nonetheless, all factors being equal, and within the resource limitations of a given environment, intelligence and power tend to be universally advantageous traits.)

These universal Darwinian drives give us a framework to talk about cancer and how to defeat it. What may be perceived as an adaptive Darwinian drive from an individual's perspective can pose a threat from the viewpoint of the colony, as illustrated in the table below, which is partially adapted from a paper by the interdisciplinary scholar C. Athena Aktipis and her colleagues. To promote multicellularity, a cell colony must devise specific countermeasures against the selfish inclinations of these Darwinian drives. This process becomes essentially a multilevel tug-of-war between two levels of selection—individual and collective.

Darwinian Drive	From the Perspective of the Individual	From the Perspective of the Colony	Countermeasure of the Colony
Replication	Replication in large quantities	Uncontrollable proliferation	Proliferation inhibition
Repair	Extension of own life span	Inappropriate survival	Controlled death
Resources	Acquisition of resources & nutrients	Resource monopoly & environmental destruction	Resource allocation & environmental protection
Power	Individual defense & offense	Internal threat to the colony	Centralized power & policing of defection
Intelligence	Differentiated strategies	Selfish differentiation	Division of labor

Let's explore this table in greater detail. First, inhibition of proliferation is key for the survival of the collective. Since the selection pressures of their environment make individual cells want to proliferate, the collective must impose strict regulation of cell division to prevent uncontrolled growth. Second, since individual cells are driven to repair themselves so as to prolong their life spans, the collective must pursue a policy of controlled cell death; this policy facilitates tissue maintenance and development while acting as a tumor suppressant. Next, because defecting cells want to monopolize resources for themselves, the collective must find ways to allocate resources efficiently, so that all cells receive the necessary nutrients and oxygen. Fourth, when rogue cells threaten to "gang up on" and crowd out other members of the collective, the colony must defend its turf by sending agents to combat defection, perhaps by producing antibodies that might neutralize defecting cells. Finally, the organism ensures a division of labor among cells by using special signals and rules written in its DNA; this stops cancer cells from doing whatever they want, and forces them to become specific cell types, such as neurons.

The exact biological explanation for how these countermeasures work is outside the scope of this book. But the broader point I want to make is that the repeated emergence of complexity across different life-forms—from algae to plants to animals—underscores a fundamental truth: the suppression of cheating is not an anomaly but, instead, a recurring theme in the history of life and human society. Darwinian angels are as endemic to life as Darwinian demons; they are the evolutionary "secret sauce" that can allow collective values to trump individual selfishness. As Charles Darwin himself wrote, "There can be no doubt that a tribe including many members who . . . were always ready to give aid to each other and to sacrifice themselves for the common good, would be victorious over other tribes; and this would be natural selection."

Constraining the Darwinian Drives

In practical terms, what can our bodies' defenses against cancer teach us about preventing defectors from taking over the entire game? We can start to answer this question by looking again at Darwinian drives—those universal adaptation strategies that, if left unchecked, can become breeding grounds for defectors and demons. Darwinian drives and their corresponding cooperative adaptations—Darwinian angels—are not confined to the biological realm. Indeed, they extend into the very fabric of human society. There are many striking parallels between biological processes, such as cancer, and certain aspects of human behavior.

Consider the use of power. In biology, individual cells may use power to proliferate at the organism's expense. Similarly, modern states, like multicellular organisms, establish governance and law enforcement to compel cooperative behavior. This centralization of power and monopoly on violence acts as a societal immune system, suppressing selfishly destructive tendencies.

It would, however, be a mistake to assume that all cooperation is angelic in nature. Dictatorships are arguably better than democracies at coercing unity and cooperation among individuals. This can create a selection pressure for coercion within institutions, effectively replacing a Darwinian demon at an individual level (for example, a violent person) with another demon at a higher unit of selection (for instance, a violent institution). In fact, I would argue that most of the Darwinian demons we have discussed thus far operate at the institutional level rather than the individual level. Indeed, democracy could be seen as a system where there is a bidirectional dependency between levels of selection: individuals can choose the rules of their institutions, and the institutions, in turn, influence the individuals.

Resource accumulation presents another parallel. Just as some cells might hoard resources, creating detrimental effects for the organism, certain actors in society seek to accumulate wealth and resources, often at the expense of others and the environment. An overaggressive logging firm, for example, might profit by directly accelerating the rate of deforestation. To counter this tendency, many modern states have implemented and enforced resource quotas and environmental protection regulations, echoing the cellular mechanisms that regulate resource allocation and maintain a balanced ecosystem.

But what exactly does it mean to "enforce" a specific countermeasure against a selfish Darwinian drive? In practical terms, any enforcement mechanism must incorporate a system of rewards or punishments, creating conditions that stop defection from being an adaptive strategy. To return to the example of the tragedy of the Grand Banks cod fishery (Chapter 1), one possible solution is to enforce fishing quotas and revoke the fishing licenses of those who don't comply. This change in dynamics can be conceptualized using a payoff matrix.

	Other Players Follow Quota	Other Players Overfish
Player Follows Quota	Sustainable fish stocks	Depletion for player, other players lose their license to fish.
Player Overfishes	Depletion for others, player loses their license to fish.	All players lose their license to fish.

In this scenario, the spread of defection is stopped dead in its tracks, due to the fact that we changed the rules of the game—i.e., the selection pressures—so that defection is no longer adaptive. However, the approach to managing such common resource dilemmas must evolve beyond simply placing individual defectors in check. While such an approach works to prevent overaggressive resource accumulation locally, it does not solve the problem on a global level. In order to truly understand and rectify the patterns of behavior that lead to collective ruin, we must investigate the structures and incentives that make defection profitable. The institutions, regulations, cultural norms, and market pressures that form the fabric of the game can inadvertently favor short-term gains over long-term sustainability, thus promoting the emergence of Darwinian demons.

It would be incorrect to assume that stable cooperation can emerge rapidly or without significant effort. The structure of human societies is built on the cooperation of individuals. These individuals are composed of cells that collaborate; cells, in turn, consist of cooperating genes. Genes themselves are produced with the assistance of the cell's metabolism, which is an intricate network of cooperating molecules. Every stratum of cooperation, from molecules to societies, has evolved over extensive periods—millions or even billions of years. This prolonged evolution is a result of the ongoing conflict between cooperative forces (Darwinian angels) and competitive forces (Darwinian demons). When stable cooperation is finally established, leading to the formation of a new unit of selection, it marks a significant event in evolutionary history, known as a *major evolutionary transition.*

The "Boring Billions"

The history of life on Earth spans about 4 billion years. For most of this time, life was predominantly single-celled, an era often referred to by geologists as the "Boring Billions." The Cambrian explosion—the sudden, initial appearance of many complex lifeforms, around 530 million years ago—marked a significant shift in this pattern. The question that perplexes many is why it took so long for complex multicellular life to emerge. In the relevant literature, this delay is often explained through environmental factors, such as moderate atmospheric oxygen levels and nutrient availability in oceans. But the concept of Darwinian demons provides another compelling lens through which to view the Boring Billions.

From its very beginning, life has been engaged in a continuous struggle between different units of selection. This struggle is characterized by the clash between opposing forces of cooperation and defection, influenced by Darwinian demons, as we've seen with cancer and the body's efforts to check its spread.

The transition from a world dominated by single-celled organisms to one teeming with multicellular life required the surmounting of numerous evolutionary hurdles. This progression depended on the successful suppression of defection (the work of Darwinian demons) and the establishment of reliable cooperative strategies (the work of Darwinian angels). When cooperative behavior became the most adaptive strategy over multiple generations, it prompted a major evolutionary transition.

Before the Cambrian explosion, life had to navigate a handful of these major evolutionary transitions. Each transition represented a victory of cooperative strategies over defective and destructive behaviors. Think of it: every breathtaking vista, every intricate ecosystem, and every marvel of nature that fills our world with wonder and awe has its roots in these epic struggles. These weren't mere skirmishes in the microscopic realm. They were monumental battles where the stakes were the future of life itself.

Every major evolutionary transition marked a momentous victory of cooperation over defection, of collective well-being over individual gain. Imagine, if you will, a world where life never evolved beyond a primordial soup of RNA molecules, endlessly replicating but never progressing. This world would be one devoid of the diversity and complexity we cherish, a world where the intricate dance of life never began.

Next, consider the momentous leap from a world dominated by single-celled organisms to one teeming with multicellular life. This transition was not just a scientific event; it was a victory over the cancerous tendencies of cellular life. Without this triumph, the world would have remained a monotonous, unvarying sea of unicellular beings.

Timeline	Unit of Selection	Defective Adaptation	Cooperative Adaptation
4.5 billion years ago	Simple molecules	**Asphaltization.** In this process, simple organic molecules such as amino acids are unable to assemble without unwanted by-products, possibly blocking complex molecules such as RNA from evolving.	**The Reverse Krebs Cycle.** This proposed model explains how simple organic molecules could have combined to form a stable cycle of chemical reactions avoiding any "asphalt" by-products.
4 billion years ago	RNA molecules	**Spiegelman's Monster.** Short RNA molecules replicate faster, thus outcompeting RNA molecules that encode useful genetic information, blocking the existence of complex genomes needed for life.	**Cell membrane.** Likely a lipid membrane, the cell membrane created a sanctuary where more complex RNA molecules could thrive, without being outcompeted by faster replicators.
3.5 billion years ago	Genes	**Selfish genetic elements.** These genes cut themselves out of one spot in the genome and insert themselves into another, ensuring their continued existence even as it disrupts the organism at large.	**Suppressor elements.** These genes suppress or police selfish genetic elements such as jumping genes.
3 billion years ago	Prokaryotes	**Viruses.** This rogue generic material tricks a cell into replicating more copies of itself while harming the host organism.	**CRISPR system.** This is the viral immune system inside cells, cutting away unwanted virus RNA before replicating.
1.6 billion years ago	Multicelled organisms	**Cancer.** Cells divide frantically against the interest of the cell colony, which may have blocked the evolution of multicellularity.	**Cancer immune system.** It inhibits cell proliferation, regulates cell death, and ensures division of labor.
150 million years ago	Groups	**Selfish behavior.** Competition for resources among individuals from the same species blocks the formation of more advanced groups.	**Eusociality.** Cooperative behavior is encoded into the genes of the first social insects, making altruism toward kin the default.
50,000 years ago	Tribes	**Tribalism.** Some tribes try to exploit and subjugate others, blocking the formation of larger cultures and societies.	**Language.** Capable of encoding norms and laws, language eventually enabled large-scale cultures and societies to form.
Now	Nations & cultures	**Global arms races.** Countries and enterprises compete for power and resources.	**?**

And then there's the human dimension. Our world, with its rich cultures, languages, and social structures, owes its existence to victories over more insidious demons—those that manifest in behaviors that seek selfish, short-term gain. Had the angels not prevailed, our world might have been a desolate place, characterized by isolation and mistrust, rather than the vibrant, interconnected web of communities and relationships we see today.

Every step forward, every evolutionary leap, has been a hard-won battle against these Darwinian demons trying to block that transition. The angels' triumphs have not just paved the way for the emergence of new life-forms; they have fostered the growth of beauty, diversity, and complexity in life. These victories are not just footnotes in the annals of biological history; they are the very chapters in the story of life on Earth. That story continues to unfold, with each of us playing a part in this grand, ongoing drama of existence.

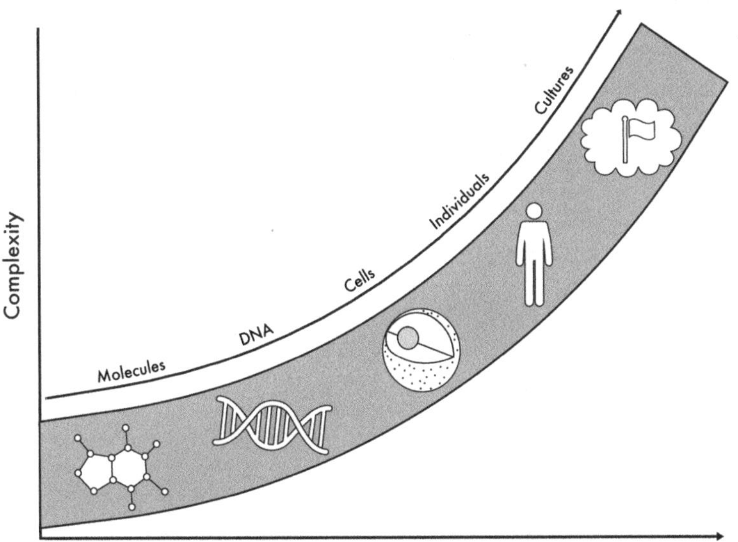

It is tempting to take the victory of these Darwinian angels for granted. After all, we like to believe that "life finds a way," to quote the skeptical mathematician from *Jurassic Park* when he explained why he believed the park's population control measures would fail. We live in an era when headlines seem to bring news of some new existential threat to life on Earth every day: global warming, viral pandemics, artificial intelligence, nuclear warfare. But if we could feel sure that life will always "find a way" to triumph against impossible odds, then those threats might no longer feel quite so scary.

Unfortunately, even a cursory study of natural history tells us that, much of the time, life does not find a way. While humans have successfully devised many strategies to promote cooperation and "find a way" on the local level, we have struggled to navigate the global-scale existential problems that might conceivably imperil the future of life on Earth. If we are to successfully solve these problems as a species, then we must abandon the deterministic notion that life will somehow always find a way to persevere, and instead actively make the sorts of cooperative choices and decisions meant to maximize our long-term odds of survival—even as we realize that Darwinian demons will be forever awaiting the next chance to strike.

CHAPTER FOUR

Demons That Threaten Life Itself

"Kamikaze Mutants"

In 2009, as part of a personal effort to become a better Python programmer, I sat down and wrote my first evolutionary simulation, in which I tried to model the interactions between predators and prey in a simplified virtual environment. Unexpectedly, though, most times I ran the simulation the predators managed to kill all the prey, thus making themselves and all other life in the simulation go extinct.

I couldn't figure out why this happened, so I delved deep into science-related web forums to determine what I was doing wrong. There I learned that, fifteen years earlier, the ecologists Hiroyuki Matsuda and Peter A. Abrams had made a similar discovery while trying to mathematically model the interactions between predator and prey. As prey adapt to better avoid predators, they inevitably spend less time eating, replicating, and doing other things that help them survive. Matsuda and Abrams discovered that these mutations could lead to the prey species eventually dying out—and taking the predator species with it. These hypothetical species of prey represented a phenomenon known as *kamikaze mutants,* so named because they engaged in what could be described as an act of evolutionary suicide.

Surely, I thought, the similarities between Matsuda and Abrams's results and my own were just a mathematical curiosity, with no real-world parallels. But the question nagged at me, especially once I learned that scientists have empirically observed the phenomenon of evolutionary suicide many times, with several different variations. Was it really true that, in certain settings, lifeforms might evolve in a way that would ensure their own eventual demise? Why do scientists keep finding examples of behavior that seems to contradict the fundamental drive for survival and reproduction that characterizes life?

When facing stressful conditions such as starvation, bacteria such as *Myxococcus xanthus* come together to share resources. Over time, though, a population of free riders has been observed to evolve, consuming group resources while sharing none, thus eventually leading to the extinction of the colony. Other bacteria have been observed to change the pH in their environments under specific conditions, in ways that ultimately led to their extinction. And then there are the northern cod, which face a selection pressure for smaller and slower-to-mature fish, since they cannot be caught as easily. However, these mutations diminished their reproductive success, since smaller cod cannot produce as many eggs as their larger counterparts. By the 1990s, the northern cod had gone commercially extinct, meaning that there were so few left that it was no longer profitable for fisheries to catch them.

In the realm of disease, certain viruses also demonstrate this principle. Viruses that are highly contagious may kill their hosts, ultimately limiting their own spread and eventually bringing about their extinction. Harmful algal blooms, too, follow this pattern; their rapid growth and the subsequent depletion of nutrients can lead to their own demise. Evidence of such self-defeating patterns in nature raises a disquieting question: Can the drive for individual advantage or survival, under certain conditions, lead not just to the downfall of a particular organism or colony but potentially to the extinction of an entire species—or even of life itself?

If so, then the premise has sobering implications for the future of humanity. The common Darwinian drives we discussed in Chapter 3—the near-universal pressures to accumulate resources, power, and intelligence—have led us to develop technologies and make choices that are fraught with potential existential peril. The drive for resources has flooded Earth's atmosphere with climate-changing carbon; the drive for power has brought about nuclear and biological weapons; the drive for intelligence has led us to create AI systems that, some fear, might ultimately supplant or destroy us. It is fair to wonder whether, by so avidly indulging these Darwinian drives, we are courting evolutionary suicide.

The idea might seem outlandish, especially when considering the resilience of life on Earth over billions of years. After all, the fact that humans are still here despite millennia worth of destabilizing choices might imply that we are not, in fact, optimizing ourselves to death. But the discovery of these phenomena in biological contexts forces us to consider the possibility that, under certain circumstances, the very mechanisms that drive human evolution and survival could also lead to our self-destruction. In order to see why this might be the case, we must first understand the concept of *survivorship bias.*

Armor-Plated Airplanes and the Apparent Invulnerability of Life on Earth

While armies and navies had been fighting on land and sea for thousands of years, airplanes were first used in combat in 1911, when the aviator Giulio Gavotti dropped bombs on Ottoman Libya in the early days of the Italo-Turkish War. Thirty years later, at the dawn of World War II, militaries were still figuring out their aerial warfare strategies—that is, how to maximize the damage inflicted on the enemy while minimizing the damage sustained by their own fleet. The pursuit of this latter goal provides us with an excellent example of the logical fallacy known as survivorship bias.

During World War II, US military statisticians were tasked with determining where to add armor to bombers that had returned from their missions. The statisticians collected data on where bullet holes were found on the returning planes. They found that the planes' wings consistently seemed to have sustained the most damage. Their initial conclusion was that the wings were the components that needed additional protection.

But upon further reflection, the statisticians realized that the data they had collected was incomplete, insofar as they had examined only the planes that had returned more or less in one piece, and had omitted data from planes that had been downed in battle. Because their dataset was taken entirely from planes that had

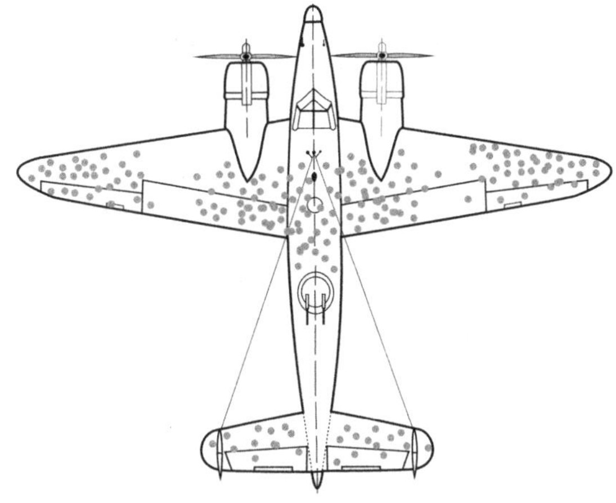

survived aerial combat, the bullet holes found in the returning planes actually indicated resilience, not weakness; they represented areas where a plane could be hit and still successfully return home. The areas in the returning planes that didn't have many bullet holes—such as the engines—were, in fact, the areas that needed more protection. Planes hit in these areas were the ones that did not return.

This example demonstrates how survivorship bias can skew our perceptions and judgments. It's a common statistical error that can generate incorrect conclusions: we focus on the entities or events that successfully passed some kind of selection or survival process but overlook those that did not. If left unchecked, this fallacy can prompt foolish and even deadly decisions.

It's easy to look at Earth, teeming with life in its myriad forms, and conclude that the organisms found here are uniquely resilient. After all, even though our planet has endured multiple mass extinction events, life has always managed to fight its way back to proliferation and prosperity. The angels of cooperation seem always to have managed to defeat the demons of defection. But is this conclusion a valid one? Or does it reek of survivorship bias?

To better understand this dynamic, let's pursue a thought experiment. Envision a colossal warehouse comprising 10 billion soundproof rooms. Each room houses an individual participating in a solitary game of Russian roulette, wholly unaware of the others' fates. With each turn, tragically, one in six players meets their end. (If we ever successfully master multiversal travel, try not to end up in the reality that contains this warehouse.)

After 126 rounds, only one participant remains. Unaware of the grim scenes that unfolded in all the other rooms, this exceptionally lucky survivor may reasonably believe that the revolver must be malfunctioning, given their continued survival amid truly daunting odds. The idea that they might lose in the next round would seem far-fetched to them, perhaps even impossible. Could Planet Earth be that lonely survivor in a game of cosmic Russian roulette?

Let's try another, similar thought experiment. There are still 10 billion soundproof rooms, but each room contains a completely closed ecosystem, with enough potatoes to sustain a single family. If they live sustainably and save a few potatoes and use them to grow even more potatoes, they can survive in there for generations. And yet one in six people born into this warehouse universe possess a "kamikaze gene," leading them to eat all the potatoes in the room, thus making it impossible to grow more of them. Tragically, with each generation, one-sixth of the rooms succumb to "evolutionary suicide."

Fast-forward 126 generations, and only a single room teeming with life remains. If you were born into the warehouse universe in the 126th generation, then this is the only room in which you could have been born. To you, whose lineage has endured for more than one hundred generations, the notion of genetic doom would appear absurd. All empirical evidence available to you would indicate that life is a stable process, with no "kamikaze genes" in sight. And yet in this warehouse universe, your mere existence offers no insight into the frequency of these genes. After all, you will always find yourself in a room that enabled your own existence—no matter how unlikely your existence might be. Instead, the truth lies in the other rooms.

Here's the lesson: examining Earth's history alone does not conclusively reveal whether life is inherently resilient, or whether it's actually very fragile and we've just been incredibly lucky all this time. Maybe we, the inhabitants of Earth, are that lone survivor, sitting in the solitary room that still echoes with life. The vacant rooms, their previous occupants silenced by insatiable mutations, represent the planets on which Darwinian demons triumphed and "kamikaze mutations" extinguished all life. Could our existence be more a product of chance than a testament to our own evolutionary invincibility? Have we willfully made ourselves blind to the fragility of life?

The Fragility of Life Hypothesis

In his 2019 paper "The Vulnerable World Hypothesis," the philosopher Nick Bostrom imagined an urn filled with balls, each representing a different technological innovation. White balls denote innovations that are entirely beneficial for humanity, such as penicillin. Gray balls represent innovations that carry some risks or drawbacks—but these risks are mostly manageable, and their drawbacks have limited impact. An automobile, for instance, might be represented by a gray ball; while emissions from fuel-powered cars are harmful to the environment, the impact of any single automobile is ultimately limited. And then there are the black balls, which signify evolutionary leaps that could spell doom for humanity; technologies and methods that would be irrevocably damaging if discovered and activated.

The same metaphor can apply to the evolutionary innovations created by natural selection. In evolutionary terms, the "white balls" might include the creation of multicellular organisms, which allowed for greater complexity and specialization in life. As we saw in Chapter 3, evolution is also rife with "gray balls," capable of blocking the development of more complex life. Examples of gray balls include viruses, predators, cancer, and parasites. While these elements might stop major evolutionary transitions from happening, on their own they are unlikely to render all life on Earth extinct. And the "black balls"? They represent the kamikaze mutations that lead to evolutionary suicide. We don't know how common they are since, by definition, we would all be dead if we ever picked one.

At its core, evolution functions as a trial-and-error process, where each mutation is analogous to randomly selecting a ball from an urn, hoping it will lead to long-term fitness. However, it gets worse. As explored in previous chapters, Darwinian demons appear to emerge invariably in competitive environments with scarce resources, such as Earth. This pattern raises a critical question: Does natural selection contain a bias toward darker shades of gray?

It doesn't require much imagination to see how some of Earth's evolutionary adaptations, if tipped to a slightly more extreme outcome, could have created a "black ball scenario." Billions of years ago, before life had a chance to diversify, all life on Earth, confined to single cells in the ocean floor's thermal vents, could have been extinguished by a lethal virus or an aggressive single-celled predator. Cancer or other selfish genetic elements might have blocked the major evolutionary transitions that make complex intelligent life possible. Bacteria and algae could have knocked out of balance the biogeochemical cycles that are crucial for life, thus preventing life from ever getting started. In all these hypothetical cases, the results are the same: a planet devoid, or nearly devoid, of complex life.

The notion that it would take only one inopportune kamikaze mutation to imperil the entirety of life on Earth forms the basis of what I term the Fragility of Life Hypothesis. In essence, this hypothesis argues that life contains the seeds of its own destruction, and that this self-defeating impulse is a tragic consequence of evolution's persistent march. Over time, evolution inevitably gives rise to Darwinian demons with the potential to obliterate all life. We are alive today only because we have been extraordinarily fortunate. For about 4 billion years, no Darwinian demon has succeeded in causing our extinction.

How likely is it that the Fragility of Life Hypothesis is true? Let's look at some of the evidence that supports the assertion that life is fragile.

EXHIBIT A:
The Fossil Record

While space travelers in some distant future may one day stumble upon extinct planets that, by virtue of their very existence, provide evidence for the Fragility of Life Hypothesis, the best we can do today is examine our own fossil record. Earth's fossil record shows that our planet has experienced several mass extinction events, sug-

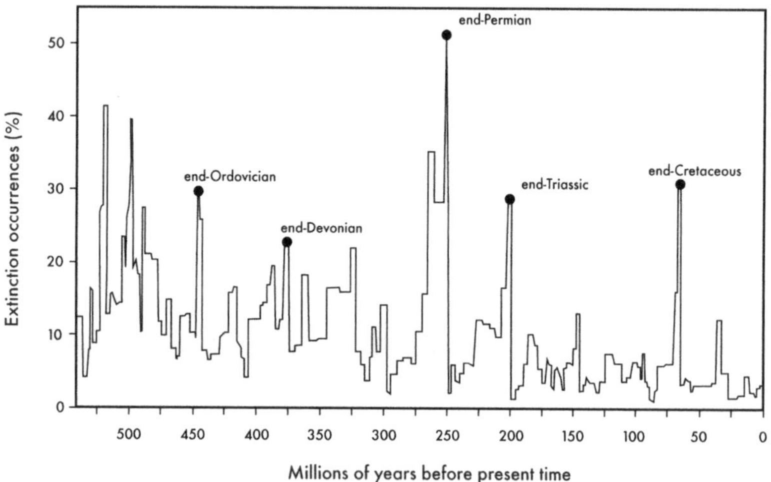

gesting that we are basically like those returning fighter jets from World War II. For billions of years, we've been lucky enough to escape with bullet holes in the wings.

The paleontologist Peter Ward, who has written extensively and convincingly on mass extinction events, suggests that a multitude of mass extinctions in our planet's history were triggered by life-forms altering the biochemistry of Earth's atmosphere, suffocating other life-forms in the process. Ward's hypothesis proposes that microbe-triggered mass extinctions will eventually result in a return to a world that is dominated by microbes and all but devoid of complex life, as has been the norm for most of Earth's history.

Ward named this hypothesis after Medea, the character from Greek mythology who kills her own children. Indeed, the Medea Hypothesis can be seen as a specific instance of the Fragility of Life Hypothesis, in which microbial mutations would lead to evolutionary suicide. Both the Medea Hypothesis and the Fragility of Life Hypothesis argue that, far from creating a utopian equilibrium, life instead tends to generate conditions that threaten its own survival.

For example, around 2.4 billion years ago, cyanobacteria began producing oxygen through photosynthesis. To many organisms at that time, oxygen was a poisonous gas. The evolutionary leap of photosynthesis dramatically altered the atmosphere of Earth, precipitating the mass extinction of many anaerobic organisms that couldn't handle the increased oxygen levels. This episode, known as the Great Oxidation Event, is just one in a series of instances when life's processes jeopardized vast numbers of species. The next table shows that the whole history of our planet has been punctuated by destructive events:

Event	Approximate Date	Hypothesized Adaptation Leading to Extinction
Great Oxidation Event	2.4 billion years ago	Emergence of cyanobacteria, which increased the oxygen concentration in the atmosphere and ocean, leading to the poisoning of organisms not used to those conditions.
Sturtian and Marinoan "Snowball Earth" Glaciations	720–635 million years ago	Evolution of more complex eukaryotic algae, which used sunlight to create oxygen, which used up a lot of CO_2, a gas that traps heat. This may have cooled Earth down.
Late Ordovician Mass Extinction	445.2 million years ago	The proliferation of simple land plants that helped break down rocks and minerals, which might have reduced the amount of CO_2 in the air. This could have led to the formation of large ice sheets.
Late Devonian Extinction	375–360 million years ago	Evolution of more complex and larger land plants, with circulatory systems. These plants could have caused soil to wash away while adding nutrients to the oceans. This led to rapid growth of algae in the oceans, which could have removed oxygen from the water.
Permian–Triassic Extinction (Great Dying)	252 million years ago	Rapid increase of bacteria that produced methane, a heat-trapping gas, after volcanic activity. This could have made Earth's climate much warmer.
Triassic–Jurassic Extinction	201 million years ago	Following volcanic activity, conditions might have become good for bacteria that produce hydrogen sulfide, a gas toxic to many life-forms.

If a large fraction of the mass extinctions we have faced were caused by life itself adapting to Darwinian selection pressures, then the Fragility of Life Hypothesis might be true. In the absence of that certainty, though, we can look outward, from the development of life here on Earth to the scarcity of it elsewhere.

EXHIBIT B:

Complex Life in the Universe Is Extremely Rare

There are billions of stars in our galaxy, many of which have been shown to have planets in their habitable zones where conditions might be right for liquid water, and possibly for life as we know it. Given the age of our galaxy (around 13 billion years), and considering that life on Earth appeared relatively quickly after the formation of our planet, life could well have emerged on other planets too. Indeed, some calculations suggest that there could be thousands, if not millions, of civilizations out there. So why haven't we encountered any?

This question is sometimes called the "Fermi Paradox," named after the physicist Enrico Fermi, who, during a 1950 conversation with colleagues about the prospects of extraterrestrial life, exclaimed, "But where is everybody?" Fermi's remark, and the paradoxical construct it inspired, reflects the apparent contradiction between the high probability of extraterrestrial life in a universe that clearly supports life, and humanity's lack of contact with or evidence for such civilizations.

Although we haven't yet discovered any extraterrestrial life, it's not for lack of trying. Over the past several decades, astronomers, scientists, and researchers have deployed myriad methods in their quest to detect extraterrestrial entities. The SETI Institute has scanned the skies for radio signals, hoping to capture transmissions from intelligent beings on distant planets. Space telescopes have surveyed thousands of exoplanets for conditions that might support life. From monitoring the atmospheres of distant planets for signs of life-supporting gases to exploring the subsurface oceans of moons in our own solar system, our efforts to find life beyond Earth have been extensive and persistent.

One common answer to the Fermi Paradox is that aliens do indeed exist; we just haven't seen them yet. According to my old Oxford colleagues Stuart Armstrong and Anders Sandberg, however, this resolution seems unlikely. In their paper "Eternity in Six

Hours: Intergalactic Spreading of Intelligent Life and Sharpening The Fermi Paradox," they posit that, in principle, an intelligent civilization should be able to spread throughout its galaxy in a relatively straightforward manner. By their calculations, a single civilization could colonize a galaxy in a few million years—a blink of an eye in cosmological terms. It is truly implausible to think that such an expansive galactic civilization would be simultaneously so secretive as to completely conceal its existence from us.

But I think that the concept of Darwinian demons offers another potential resolution to the Fermi Paradox. These universal selection pressures, which render life susceptible to self-destruction, might have extinguished extraterrestrial life on every planet where it was once found. Perhaps even more likely, life never got that far: Darwinian demons might have prevented major evolutionary transitions everywhere else but here, thereby inhibiting the emergence of complex life capable of broadcasting its presence across the universe. In a paper titled "The Timing of Evolutionary Transitions Suggests Intelligent Life Is Rare," Andrew Snyder-Beattie, Anders Sandberg, Eric Drexler, and Michael Bonsall argued that the major evolutionary transitions required to produce intelligent life are so improbable that we might be the only intelligent beings in our observable universe. In this scenario, Fermi's Paradox perhaps isn't a paradox at all, but an elegy. Where is everybody? Look around: We might be all we've got.

EXHIBIT C:

Life Is Constantly on the Precipice

Many scientists agree that climate change, nuclear war, bio-risks such as man-made pandemics, and the uncontrolled development of artificial intelligence present significant threats to life on this planet. These scenarios all emerge from the Darwinian drives discussed in Chapter 3. To survive, any agent must secure resources. This pursuit of resources has caused environmental harm on a global scale, potentially setting the stage for a new mass extinction. In a competitive world with finite resources, an agent must become powerful to access land and resources—a dynamic that has led to the creation of nuclear weapons. These weapons, if used in a global conflict, have the potential to destroy most, if not all, of human life. To effectively wield power and acquire resources, an agent needs intelligence, the pursuit of which has driven the development of artificial intelligence. However, if left unregulated, AI could become uncontrollable, resulting in dire consequences for humanity.

These scenarios aren't merely speculative, science-fictional ones; they are clear and present dangers that we are grappling with right now. If the Fragility of Life Hypothesis holds true, then these looming threats are not aberrations or the result of bad luck but the inevitable consequence of the evolutionary process that has, paradoxically, allowed us to thrive. We may have won numerous previous battles against Darwinian demons, but the war is far from over.

So what recourse do we have? One idea that occasionally resurfaces in academic discourse is that life on Earth would be better off if humans didn't exist—that we've simply become too destructive a force. For instance, the Voluntary Human Extinction Movement is an environmental campaign that calls for all people to abstain from reproduction so that we may gradually go extinct. I disagree with this standpoint, which strikes me as a rather extreme overcorrection to a fundamentally misdiagnosed problem. If the Fragility of Life Hypothesis is correct, then it's nature itself, rather than humanity uniquely, that has a predisposition to self-destruction.

As a result, we find ourselves in a singular position. After about 4 billion years of cosmic serendipity, we may very well be the only form of life in the universe capable of escaping the tyranny of Darwinian demons. Unlike cancer or viruses, forged by the trial and error of natural selection, we have the chance to intentionally guide our future development according to principles of rationality, empathy, and long-term planning. This aspect of humanism may be not just our own saving grace, but the saving grace of life itself.

But the path to successfully defeating Darwinian demons is fraught with dangers. Some have suggested that humanity is akin to a heedless driver steering a car directly toward the lip of a cliff. Considering the rise of exponential technologies, one might modify this analogy by suggesting that the cliff-bound jalopy humanity is driving has been magically swapped with a turbocharged sports car. We're heading toward the same cliff—but we're speeding toward it faster than ever before.

The rise of exponential technologies, such as AI, which require ever shorter periods of time to grow and develop and change our world, means that now is the time to defeat our kamikaze mutants before they destroy us all. We are very plausibly the last inhabitants of our universe. Nothing inherently special about us guarantees our survival, and it would be unwise to presume otherwise. More likely, we've just been really lucky all this time—and if we don't take active steps to reset the incentives, ward off disaster, and safeguard our future, then our luck will inevitably run out.

This sobering prospect is what we'll explore in Part II.

PART II

At the Precipice

CHAPTER FIVE

The Depleted World: The Resources Arms Race

In the Prologue, we toured the crumbling remains of Picher, Oklahoma: a once-thriving mining settlement that is now completely abandoned. Aggressive overmining in the region created by-products so toxic that the rivers ran red with contaminants and the entire town became uninhabitable. But before Picher became a full-fledged ghost town, it was briefly home to some of the most valuable land in the United States.

In the first half of the twentieth century, the lead and zinc deposits buried in the hills of northeastern Oklahoma were critical to global security and domestic prosperity. The mining infrastructure that sprang up around Picher was meant to extract those minerals as quickly and efficiently as possible, so that they could be used in a wide range of products. And for a time, the world put Picher's resources to very good use—even as the act of extracting them doomed the town to extinction.

You can't really blame the Picher mining companies for taking advantage of the minerals buried underground there. After all, the continued existence of life on Earth requires life-forms to make use of their habitats in one way or another. Organisms metabolize energy and materials to sustain themselves; communities fell trees and quarry stones to build houses and other structures; factories take raw materials and turn them into useful goods. There is nothing inherently wrong with extracting and consuming resources in this manner.

Things don't get dicey until Darwinian selection pressures enter the equation—specifically, the pressure to extract and consume resources at harmful and unsustainable levels. This rapacity is often a function of increased efficiency. A person who learns to climb a tree and pick the bananas hanging from its high branches will be better equipped for survival than those who settle for trying to dislodge the fruit by throwing rocks at it from the ground. A mining company that uses dynamite to extract minerals from the earth will be more productive and thus more profitable than one that relies exclusively on miners with pickaxes. Humans first burned dung for fuel, then wood, then coal, then oil; with each transition, we got more and more utility out of the fuels we burned.

This selection pressure for increased efficiency is what motivates the Darwinian drive to accumulate resources. This drive, in turn, has brought about a *resources arms race*. Resources might take the form of minerals or nutrients; they might be money or securities or land. By getting better and better at extracting and using them, individuals and entities hope to increase their own odds of survival—even if doing so diminishes those odds for everyone else.

The effect of this arms race on a given environment isn't necessarily a matter of gluttony, where one person's piggishness at the table leaves less food for all the other diners. Instead, these trends toward increased efficiency can encourage patterns of aggressive overexploitation that, if left unchecked, can destabilize the surrounding environment. The players in a competitive resources arms race rarely start out with destructive intentions, and the harm they may inflict on their environment is generally a by-product of their methods, not the goal of those methods. The first people to burn oil for fuel, for example, surely did not realize that the fumes emitted by that process would one day come to wreak environmental havoc. Likewise, while the zinc executives of Picher, Oklahoma, surely knew that mining was inherently unhealthy work, they probably did not think that the chat piles heaped up around town would eventually drive the town to extinction.

Yet the pernicious thing about an arms race is that by the time a player realizes the widespread harm it has wrought, it can feel difficult or impossible to opt out of the race and choose a different path. Imagine that you're the proprietor of a Grand Banks cod fishery. Like your competitors, you have chosen to overfish your shared waters in order to bolster the short-term survival odds of your business. You know this practice will eventually spoil the waters and doom all the cod fisheries, including your own—but what recourse do you alone really have in the short term? If you make a principled stand and decide to abide by the quota even as your competitors keep overfishing, then yours will be the first fishery to go out of business, and your sacrifice won't stop the waters from getting fished out anyway. And so you trudge on toward the finish line of your resources arms race, even as you know that it will end only in catastrophe.

Nick Bostrom's metaphor of an urn filled with white, gray, and black balls can help us understand the escalating scale of resource-related catastrophe. In this context, the color gradient of the balls corresponds to increasingly efficient methods of extracting and using resources. The darker the ball, the graver the macroscopic consequences of defection—of exploiting resources at an unsustainable rate.

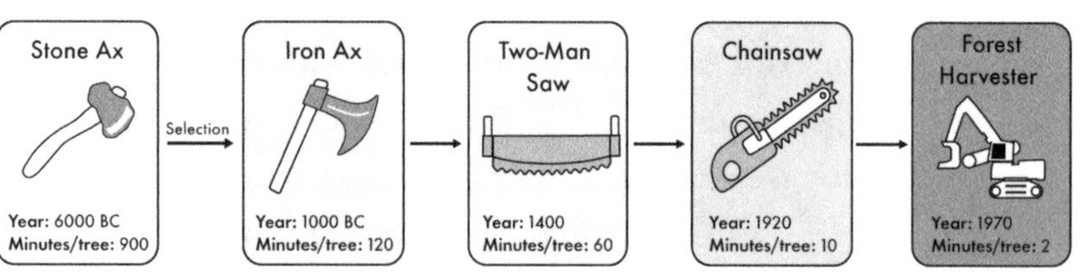

Mining minerals with a pickax, for example, is a fairly low-impact method of resource extraction and would likely correspond to a whiter-hued ball. The individual miner might become ill because of his labors, but those labors are unlikely to directly imperil anyone else. A high-pressure milling process is certainly a more effective means of separating metals from ore, but the by-products are also more consequential: toxic mine tailings can seep into the groundwater or be carried away by the wind.

To move toward a more contemporary example, fossil fuels power the infrastructure of modern life—but burning those fuels spews carbon dioxide into the atmosphere, which in turn traps heat and causes global temperatures to rise. Similarly, improved technologies for forest harvesting, from stone axes to modern harvesters, have led to 10 percent of the world's forests being cut down since 1900. This rampant deforestation has had profound global consequences: increased soil erosion makes land less fertile for agriculture, the loss of tree cover negatively affects biodiversity, and fewer trees to convert CO_2 into oxygen necessarily accelerates the pace of global warming.

As implied by my earlier invocation of the Grand Banks cod fishery, the resources arms race is a classic example of the tragedy of the commons, in which individual actors are free to draw from a common pool of resources, but no single actor claims responsibility for the continued welfare of that common pool. As humans get more and more efficient at extracting and using resources, they tend to inflict more and more damage on the commons.

Many of today's modern corporations have taken the resources arms race to new levels. These companies are often legally obligated to attempt to maximize shareholder value—a requirement that can make aggressive resource exploitation feel like a moral imperative. The broader consequences of this monomaniacal focus on profit have been disastrous for our biome, most notably in the ongoing effects of anthropogenic climate change.

Our world is remarkably resilient, but its resilience is also a function of ecological balance. The implicit bargain between Earth and its inhabitants has always been that we must exploit its resources at a sustainable rate. Today, though, not only are we going through resources at an unsustainable rate, we are doing so in a manner that imperils the future prospects of life on this planet. Before we can attempt to fix this problem, we must first understand the ways it has played out in the past and present—and we must understand that the story of the resources arms race is also the story of a concept called *metabolism*.

Metabolism: The Invisible Killer

Life cannot exist without drawing on its environment. It requires specific resource inputs—including materials, minerals, and nutrients—in order to move, live, survive, and reproduce. Metabolism, or the act of converting nutrients into energy, is a process vital to every life-form. But this process also creates metabolic waste, an inevitable consequence of the consumption of resources. In fact, just by breathing, you have an impact on the

Circular metabolism of nature

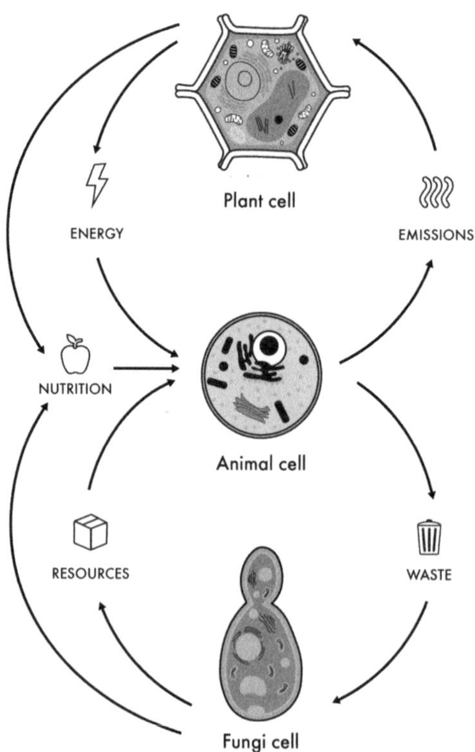

environment. Inhaling oxygen causes you to exhale carbon dioxide, a greenhouse gas that, if emitted in sufficient quantities, can warm the globe to unsustainable levels.

For life to thrive and survive, the metabolic processes must balance each other out, in a cycle sometimes called *circular metabolism*. The metabolic by-product of animals breathing must be counterbalanced by photosynthesis from plants. Likewise, without decomposers such as fungi, bacteria, and worms, the accumulation of organic waste would disrupt ecological systems. Combined with geological processes, these organisms form planet-wide chemical cycles, in which chemicals are repeatedly circulated between other compounds, states, and materials, and then back to their original state. Forget what you learned from *The Lion King*—this is the *real* "circle of life."

But as I implied in Chapter 4 when discussing the Medea Hypothesis, systems rarely boast a perfect balance. More often, their metabolic processes are imbalanced, which serves to destabilize those systems. If the circumstances are right, an imbalanced metabolism might even engineer the total annihilation of other species, via resource depletion, waste accumulation, and environmental pollution.

You could also say that cities have their own metabolisms. Resources such as water, food, raw materials, and energy enter the urban system, get processed in various ways—in factories, homes, businesses, and transportation systems—and then leave the system as waste, emissions, and heat. This intricate process, much like the metabolism of a living organism, is essential if cities are to develop and thrive. But at the dawn of the Industrial Revolution, the process began to tip substantially into imbalance.

The Industrial Revolution—running from roughly 1760 to 1840, and then again from 1870 to 1914—was arguably the most efficient prosperity engine that humans had yet known. Not only did the advent of factory production and standardized processes allow us to turn raw materials into a wider array of useful goods than we'd ever had before—which, over time, raised baseline standards of living in all industrialized nations—it also created wealth, in differing proportions, for industrialists and workers alike. (This is not to say that the life of a nineteenth-century factory worker was a good one; it's just to say that the Industrial Revolution transformed their economic relationship to their environments.) This transition helped to urbanize societies, bringing workers off the farms and into the places where the factories were, thus catalyzing the growth of great cities.

The Industrial Revolution was itself a metabolic process, one that brought the modern world to life. But the Industrial Revolution also created significant amounts of metabolic waste, everything from unhealthy working and living conditions to waterways tainted with sewage and factory by-products. The most immediately visible form of metabolic waste was the coal smog, emanating from the chimneys and smokestacks of homes and factories, that came to blanket industrial cities. This smog was most associated with London and other British cities, which were at the forefront of the Industrial Revolution; the term "London fog" initially referred to smog so dense that it could almost be held in your hand. The London fog was dangerous to humans, animals, plant life, and buildings alike—an inadvertent and yet seemingly inescapable cost of creating the modern city.

These days the repercussions of an imbalanced urban metabolism are rarely limited to specific cities and regions. Today's metropolises are so big, and their metabolic processes so intensive, that the consequences of imbalance can extend far beyond their

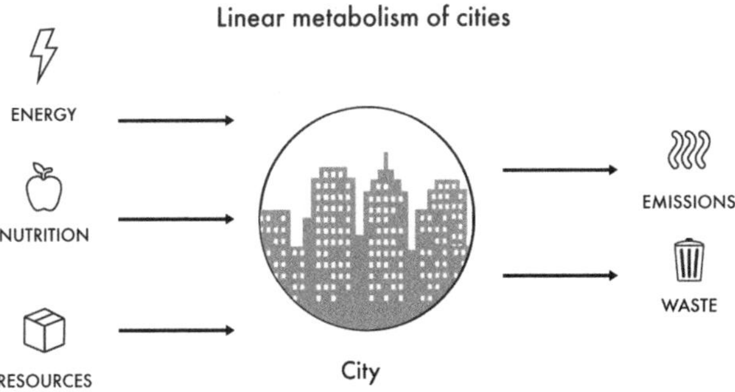

city limits. These consequences can be summed up under two broad categories: *limits on raw materials* and *strains on planetary boundaries.*

The Problem of Overconsumption

The urban metabolism that fuels our societies is a powerful force, incessantly requiring more resources as our cities continue to expand. But these demands inevitably run up against the undeniable truth that our planet's raw materials are finite. We are consuming them faster than they can be replenished, and thus every extraction brings us closer to their eventual depletion.

Take, for instance, oil. Black gold powers much of our world. Experts have long debated when we will reach "peak oil," or the point at which global oil production reaches its maximum rate before it begins to decline. Current estimates suggest that we could reach this point within the next two decades. Crossing this Rubicon will have severe implications for our energy-dependent society.

Many other crucial resources, from copper to phosphorus, are also under threat. Their depletion will pose significant challenges to the many industries, technologies, and economies that rely heavily on these materials, causing everything from widespread job loss to the demise of entire sectors. Copper, for instance, is an extremely conductive metal that's integral to a wide range of devices, from electrical wiring and batteries to smartphones and computers. Reaching "peak copper" would, among other things, directly affect the maintenance and expansion of electrical grids; impede the production and adoption of electric vehicles, thus contributing to increased global warming; slow the pace of technological progress while affecting worldwide telecommunications networks; and create countless second-order consequences.

The extraction processes for these crucial resources often lead to environmental destruction and pollution, further exacerbating the broader problem of metabolic imbalance. Take the island of Borneo, which for millennia was dense with tropical rainforests. In the final decades of the twentieth century, aggressive deforestation performed by logging companies destroyed more than half of Borneo's rainforest. Much of the newly bare land was repurposed for agriculture, but the deforestation had diminished soil quality, negatively impacting the land's fertility. The logging process affected the health of the remaining trees, meanwhile, leading to frequent and debilitating wildfires. In the span of fifty years, the resources arms race has taken Borneo's rainforests and replaced them with scenes of environmental destruction.

In fairness, I should note that people have been predicting the end of resources for a very long time, and their predictions have not always been successful ones. The graph on page 95, a version of which was originally published by the magazine *New Scientist* in 2009, predicted that indium—a metallic element used in flat-panel displays—would be depleted by 2022 if the world did not

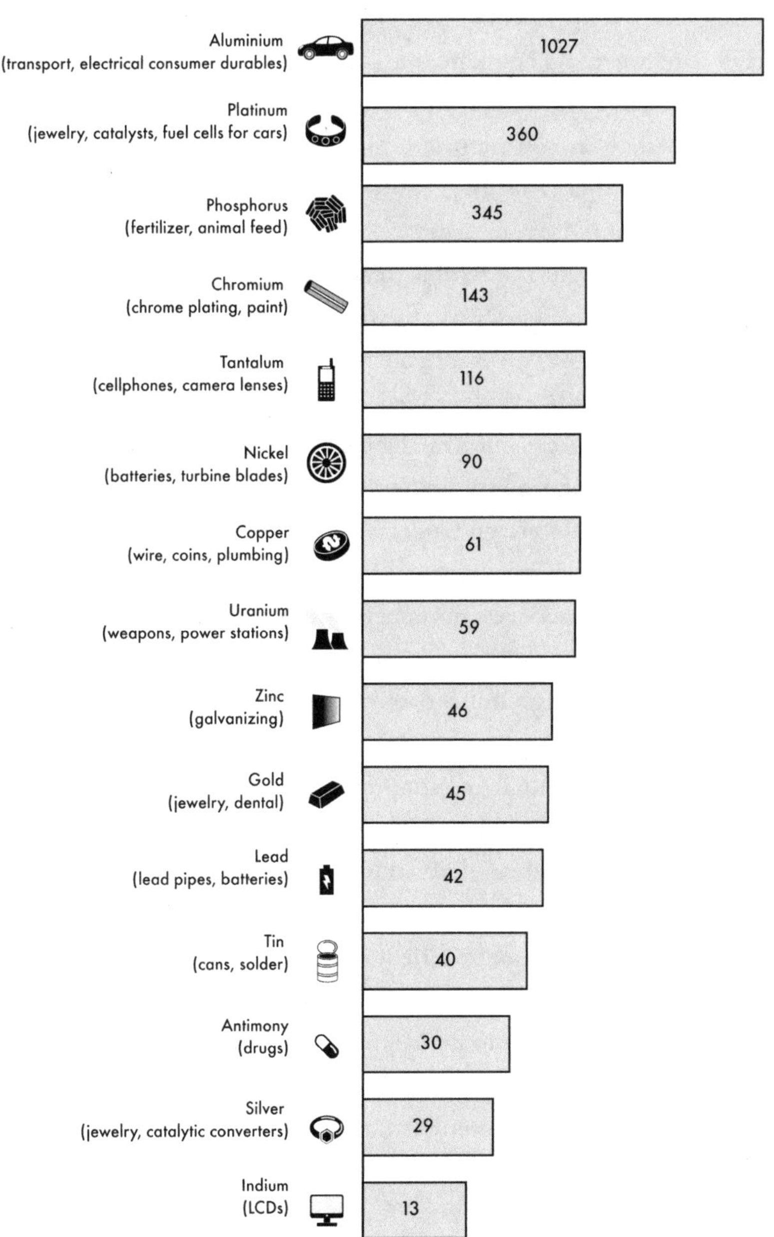
Years Left of Resource
(Y-axis not at scale)
Aluminium
(transport, electrical consumer durables)
1027
Platinum
(jewelry, catalysts, fuel cells for cars)
360
Phosphorus
(fertilizer, animal feed)
345
Chromium
(chrome plating, paint)
143
Tantalum
(cellphones, camera lenses)
116
Nickel
(batteries, turbine blades)
90
Copper
(wire, coins, plumbing)
61
Uranium
(weapons, power stations)
59
Zinc
(galvanizing)
46
Gold
(jewelry, dental)
45
Lead
(lead pipes, batteries)
42
Tin
(cans, solder)
40
Antimony
(drugs)
30
Silver
(jewelry, catalytic converters)
29
Indium
(LCDs)
13

drastically curtail its rate of consumption. This prediction obviously didn't come true; two years after the graph's proposed depletion date, we're still cranking out LCD screens. Whenever we start to run out of a resource, the price of that resource goes up, which tends to slow its consumption. Simultaneously, the incentive to find more of that resource gets stronger, and when more is found, prices tend to normalize again.

While we might not be able to accurately predict when we will fully deplete a given resource, the fact remains that the Earth does have limits, and we will eventually run out of some of the resources on which humans have depended for so long. Before we fully deplete some resources, they might reach a state of effective depletion, where their scarcity and/or inaccessibility makes them so expensive that they are no longer worth extracting.

Another crucial component of our consumption patterns involves ecosystem services: the benefits humans gain from nature. A healthy natural environment positively affects climate and water quality while having a mitigating effect on disease and certain natural disasters, benefiting all of us. Nature also provides recreational, aesthetic, and spiritual benefits that, while perhaps difficult to quantify, are inherently important to human well-being. And we mustn't forget that none of us would exist today were it not for the role that nature plays in life-supporting services such as soil formation, photosynthesis, and nutrient cycling.

But our relentless consumption is putting a strain on these services as well. Our oceans are overfished, our forests depleted, and our soil eroded. These issues are not just environmental ones; they directly impact human life. For example, the total collapse of the world's fisheries would be more than just a loss for biodiversity—it would threaten the food security of billions of people who rely on fish as a primary source of protein.

While forecasts vary as to when exactly we might run out of crucial resources or reach peak ecosystem services, it seems evident that we are eating into our natural capital at an alarming rate, and the trend is simply unsustainable in the long run. The gravity of the situation becomes even more pronounced when we consider the *planetary boundaries* that mark the safe operating space for humanity.

While resource depletion would make life on Earth more difficult and unpleasant for its many inhabitants, running out of copper, for example, would not on its own cause life itself to completely collapse. But it would be a different story entirely if we were to transgress Earth's planetary boundaries. Overshooting these boundaries could create even more severe consequences for our societies and the natural world—consequences that would be catastrophic for the entire planet.

Breaching the "Safe Zone" for Life

Imagine that it's a crisp winter morning and you've decided to go outside and build a snowman. For a few days afterward, the snowman stands tall in your front yard—until the weather turns unexpectedly warm. This unseasonable weather is bad news for your snowman, which cannot survive once the temperature rises above freezing. Over the next few days, the snowman melts and shrinks until there's nothing left but a few lumps of coal on the grass.

This familiar example offers an easy way to understand the concept of *planetary boundaries*—a scientific framework that outlines the viable operating space for humanity with respect to Earth. Just as a snowman can survive only within a specific temperature range, these planetary boundaries constitute the "safe zone" for life. As our urban metabolism ramps up, we get closer to and in some cases exceed these critical thresholds, thus risking irreversible and potentially catastrophic environmental change.

There are nine recognized planetary boundaries, the most commonly understood of which is *climate change*. While the other eight boundaries are perhaps not as well known, the consequences of breaching them would also be dire. One of these boundaries is the *loss of biosphere integrity*, in which species go extinct and biodiversity suffers as a result. The others are: *stratospheric ozone depletion*; *ocean acidification*; *biogeochemical flows*, such as phosphorus and nitrogen cycles; *land-system change*, such as deforestation; *freshwater change*; *atmospheric aerosol loading*, which refers to microscopic particles in the atmosphere that affect climate and living organisms; and the *introduction of novel entities* into earthly habitats, such as organic pollutants, radioactive materials, nanomaterials, and microplastics.

Scientific research indicates that we've already exceeded the safe limits for climate change, biosphere integrity, and biogeochemical flows. Our global thermostat is being cranked up rapidly, with carbon dioxide levels now higher than they've been at any point in the last 800,000 years. The rate of species extinction is tens to hundreds of times higher than the average rate in prehuman times, and our manipulation of the phosphorus and nitrogen cycles is so profound that we've exceeded that particular planetary boundary by 100 to 200 percent.

The implications of crossing these boundaries are grave, potentially leading to new, less stable states of Earth. In a future where

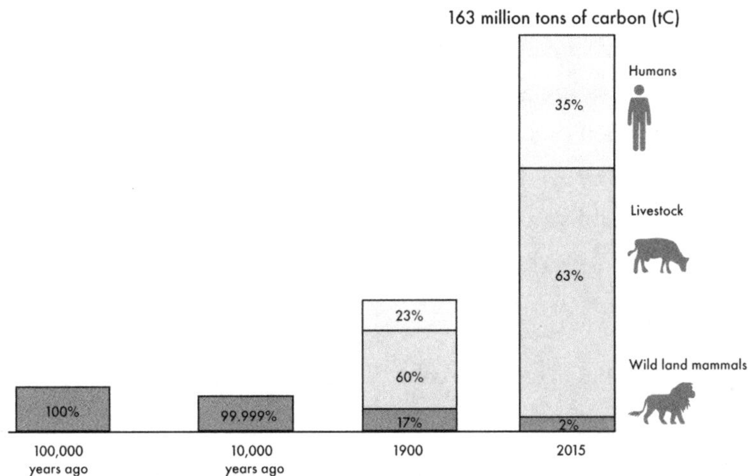

humanity has crossed all planetary boundaries, Earth is transformed into a world starkly different from the one we know today. Climate change has escalated unchecked, leading to extreme weather events: hurricanes are more powerful, droughts longer and more severe, and wildfires more frequent and devastating. Ice caps have melted, causing sea levels to rise and submerge coastal cities, displacing millions. Biodiversity has suffered a catastrophic decline; numerous species have been driven to extinction, disrupting ecosystems and the services they provide, such as pollination, which is essential for food production. Air and water quality have deteriorated significantly, leading to health crises worldwide. Food security is severely compromised as agricultural systems struggle to adapt to the changing climate, leading to widespread famine. Societies face mounting pressures as they grapple with the challenges of displacement, resource scarcity, and the political turmoil that follows.

While our planet is resilient, we should not conclude that it will thus always be hospitable to all forms of life—including human life. Our fundamental challenge, then, must be to steer away from these boundaries and guide our urban metabolism to exist within the safe operating space. Much like natural ecosystems—which have developed cyclic metabolisms to recycle carbon, nitrogen, oxygen, phosphorus, water, and sulfur among organisms—humanity too must establish a circular economy.

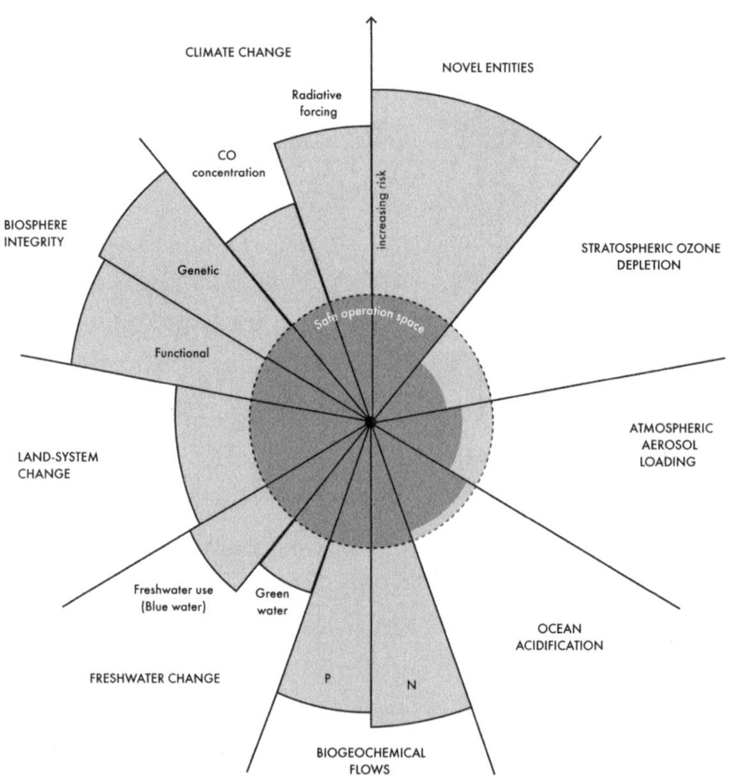

In a circular economy, materials within our urban metabolism would be continually reused, fostering a sustainable equilibrium. This approach would mimic the efficiency of nature, where nothing is wasted and everything serves a purpose in a perpetual cycle of use and reuse. Adopting such a system will be essential for maintaining balance and sustainability in our increasingly urbanized world.

The Darwinian Traps of the Resources Arms Race

It should not be controversial to say that we've exceeded the limits of our planet, depleting essential resources and witnessing widespread species extinctions. So why aren't we taking action to fix the problem and restore ecological balance? The answer, once again, comes down to Darwin's drives and demons.

Many scientists and activists have tried to take meaningful climate action. The 1972 United Nations Conference on the Human Environment in Stockholm was the first major international conference on environmental issues, leading to the creation of the United Nations Environment Program (UNEP), with a broad mandate to coordinate on environmental issues within the UN. Other efforts followed, including the United States' 1973 Endangered Species Act, which provided extensive protection for endangered species; the 1987 Montreal Protocol, which successfully sought to address the depletion of the ozone layer; the 2015 Paris Climate Agreement, which commits signees to keep the rise in mean global temperature to well below 2°C (3.6°F) above preindustrial levels; and, more recently, the COP28 conference in Dubai, where nearly two hundred countries agreed to "transition away" from fossil fuels.

These efforts have had some positive effects on the rate of climate dissolution. Many countries have managed to substantially reduce their emissions; many have found ways to decouple economic growth from carbon emissions. And yet despite these valiant efforts, our planet continues to warm, and companies continue to exploit their environments in pursuit of profit. Why haven't the aforementioned acts and treaties been effective enough to solve the problems they're meant to solve? The answers lie in the selection pressures on politicians and corporate leaders—and the fact that, for them, it is often adaptive to prioritize short-term gains over long-term sustainability.

As we discussed in Chapter 2, politicians face a constant pressure to optimize for electability. In a legislative context, this pressure leads them to focus on immediate popular issues such as job creation, economic growth, immigration, and taxation. This short-term focus often sidelines long-term environmental concerns. A significant example is the withdrawal of countries from international environmental agreements. Notably, in 2017 the United States, under the administration of President Donald Trump, announced its withdrawal from the Paris Agreement, citing economic burdens and unfair advantages to other countries as reasons. (The United States has since rejoined the Paris Agreement.)

Similarly, in 2011 Canada withdrew from the Kyoto Protocol, the treaty adopted in 1997 that set internationally binding emission reduction targets for developed countries; then-environment minister Peter Kent stated that meeting the Kyoto targets would cause significant economic losses. Canada has since signed the Paris Agreement, which builds upon and significantly expands the scope of the Kyoto Protocol. And yet in a world where nations' decisions to respect or reject international environmental agreements are motivated by immediate political and economic considerations more so than questions of long-term environmental impact, there is every reason to think that similar withdrawals can and will happen again.

Likewise, in the corporate world, the selection pressure for profit maximization has led to decades of inaction at best and stark opposition to any tangible action at worst. Major oil companies, despite being aware of the impacts of climate change as early as the 1970s, have spent millions on lobbying and funding misinformation campaigns. In 1977, for instance, scientists working for ExxonMobil, one of the world's largest oil companies, informed company executives that fossil fuels contributed to climate change. Yet between 1998 and 2014, ExxonMobil spent nearly $31 million on organizations that promoted climate denial. These efforts significantly influenced government policies: for example, when the United States decided not to ratify the Kyoto Protocol in 2001, its hesitation was in part thanks to the lobbying of the fossil fuel industry.

Here's another major issue: it is often easier and more cost-effective for companies to present an eco-friendly image than to implement real, substantive change. *Greenwashing* is what happens when companies make misleading claims about the environmental benefits of their products or practices. In 2015, in perhaps one of the most infamous cases of greenwashing, the automaker Volkswagen was found to have installed software in its diesel vehicles that could detect when they were being tested for emissions. This "defeat device" software manipulated the vehicles' performance during regulatory testing so that the vehicles would meet US standards. But when those vehicles were driven in the real world, they emitted up to forty times more nitrogen oxide than they did during testing. Volkswagen had marketed these vehicles as environmentally friendly and low-emission, deceiving both consumers and regulators. The scandal is a practical example of Goodhart's Law in action. Although modern companies routinely set climate targets for themselves, they also routinely fall short of those targets, because financial markets continue to select for companies that produce the most profit.

The market for carbon offsets also illustrates how enterprise profit incentives can undermine environmental goals. You may have seen the label "carbon neutral" on some products in supermarkets, or you may have been asked if you want to "offset" your emissions when you buy a flight. In order to compensate for the emissions they produce, companies invest in planting trees that should remove an equivalent amount of carbon from the atmosphere.

In theory, the idea of carbon offsets sounds great. In practice, it has two serious flaws. First, companies do not always account for their carbon emissions entirely accurately. A recent working paper titled "Emissions Gaming?" highlights that the carbon emissions of some enterprises could actually be several times higher than those currently reported, due to ambiguities in the ways carbon emissions are accounted for.

Second, once companies have calculated their emissions, they often buy low-quality offsets as a way to claim carbon neutrality without making substantial reductions in their own emissions. This practice would not be too bad if these offsets did what they promised. But a recent study, involving twenty-six projects claiming a total of 89 million carbon credits, found that over 90 percent of them likely failed at avoiding emissions entirely. This is no niche industry: these credits were akin to the combined annual emissions of Greece and Switzerland.

Unfortunately, these two shortcomings can have a multiplicative effect. For example, certain businesses might report only 10 percent of their actual emissions, then invest in carbon offsets. However, if only 10 percent of these offsets effectively prevent emissions, the result is that merely 1 percent of the emissions are genuinely mitigated. Yet these businesses might still label themselves "carbon neutral."

Accurate carbon accounting is more than a technical necessity; it's the backbone of environmental accountability. This belief motivated me to co-found Normative, a carbon accounting software company, in 2014. The goal was to empower companies to accurately account for their carbon emissions. Over the past decade, I have advocated for the implementation of mandatory, standardized, and precise carbon disclosures. Alongside this advocacy, I've pushed for carbon taxes that would not only enable but also compel enterprises to recognize and internalize carbon emissions as an external cost.

Throughout my career, I've encountered numerous business leaders and politicians who have demonstrated remarkable courage in their efforts to address climate change. However, the need to label these actions as "courageous" is a clear indication of a systemic flaw. To truly combat climate change, we shouldn't have to rely on individual acts of bravery. Instead, doing the right thing should be the rational choice, which in turn would mean creating a system where the interests of businesses and politicians naturally align with environmental stewardship. Creating systemic change shouldn't just be about encouraging courageous decisions; it must also involve reshaping our economic and political frameworks to make sustainable choices about the most logical and beneficial path forward. It must be about slaying the Darwinian demons that make us act selfishly in the first place.

Domination of the Finite

As the guardians of our own future, we must reckon with these Darwinian demons—not with fear, but with the resolve to redefine our relationship with the world that sustains us. If we fail to do so, then we risk our future becoming a series of escalating depletions until life on Earth finally gives its last gasp. I began this chapter by returning to the story of Picher, Oklahoma—the once-vibrant mining settlement that was literally exploited into extinction. If the world does not learn from the lessons of Picher, then it risks becoming *like* Picher. Humans are an intelligent species capable of learning from their mistakes. We can choose to avoid turning Earth into a global ghost town.

The need to respect and work within our planetary boundaries has never been more critical. Imagine a future where the relentless pursuit of resources escalates into an even more perilous competition. In a world constrained by finite resources, the stage is set for a relentless global contest, where the mightiest armies often claim victory. History tells us that almost every armed conflict has been a fierce battle for precious resources—fertile land, rare minerals, clean water, or oil. The rule of the game has been simple: the side with the most formidable military force usually emerges triumphant.

The last few decades have been largely peaceful ones. The world has managed to attain exponential economic growth by overexploiting nature's limited resources. But without a shift toward a sustainable and circular economy, one that honors our planet's natural limits in a closed-loop urban metabolism, we risk not just continuing this dangerous cycle but intensifying it. In fact, each standard deviation increase in temperatures is increasing the frequency of armed conflict by 14 percent. Armed with weapons of ever-greater destructiveness, jostling to control the remaining resources in an increasingly depleted world, we have come to fuel a vicious power arms race.

CHAPTER SIX

Extinction Weapons: The Power Arms Race

In 1492 three ships set off from Spain on an exploratory voyage, hoping to find a new maritime trade route between western Europe and Asia. Instead, they washed up in the Bahamas, an uncharted land filled, as expedition leader Christopher Columbus later assured his funders, with untold riches. If Columbus overestimated the amount of gold to be found in the Bahamas, he assuredly underestimated the scope of the opportunity that the discovery of the New World presented for the entrenched powers of the old one. More ships from Europe soon followed. The European colonization of the Americas had begun.

Until very recently, generations of Western schoolchildren were taught to view Europe's arrival in the Americas as an unmitigated triumph, a turning point in world history wherein civilization and culture took root in a primitive realm. But for the indigenous people of the so-called New World, Columbus, Cortés, and the other explorers who followed weren't settlers so much as conquerors—

imperious interlopers who sought to seize the native peoples' ancestral lands.

When multiple parties lay claim to the same plot of land, there are only a few plausible resolutions. In the civil realm, such disputes are generally resolved via mediation, compromise, or monetary exchange. In the annals of empire, they have most often been resolved by force. The use of force is an adaptive response to the environmental selection pressure to increase power—a universal Darwinian drive that is generally deployed in pursuit of additional resources. This selection pressure is a prototypical Darwinian demon, encouraging adaptive behaviors that can bring short-term gain to those who increase their power, while wreaking havoc on everyone else.

When power is wielded in organized fashion by nations and empires, it can quickly become an *existential* demon, one that can lay waste to entire populations and civilizations. Since the use of power generally begets subsequent uses of power, this demon can easily spark a *power arms race,* in which nations and empires hasten to develop better weapons and military strategies, so as to ensure their own survival while imperiling their enemies. With each new development in the use of power, another ball is plucked from Bostrom's urn.

And each subsequent ball seems to have a darker and darker shade.

The weapons and tactics used by European colonizers may not have been powerful enough to endanger the entire world, but they nevertheless laid waste to the people and civilizations of the Americas. Within two hundred years of first contact, European settlers had decimated indigenous populations there; by some estimates, native population numbers dropped by as much as 80 percent. In

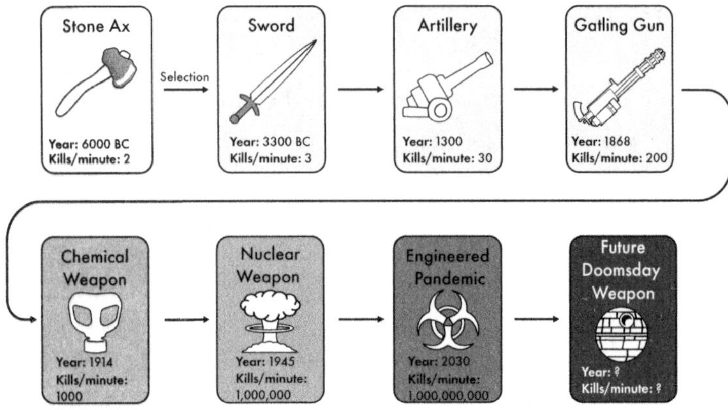

a dramatic clash of civilizations, Hernán Cortés and Francisco Pizarro, leading mere handfuls of Spanish conquistadors—600 for Cortés and 170 for Pizarro—toppled the mighty Aztec and Inca empires with astonishing speed, a feat largely attributed to their advanced military technology.

These conquerors were armed with the gleaming steel of swords and armor, groundbreaking firearms, and the imposing presence of horses—elements that were entirely alien and terrifying to the indigenous warriors. The Aztec and Inca, in stark contrast, wielded weapons crafted from wood, stone, and bronze. These obsidian-bladed clubs, spears tipped with the same volcanic glass, and simple slings and bows were effective in their way, but they could not compete with the advanced technology wielded by the Europeans. (Smallpox brought from Europe further devastated the Aztec and Inca societies, weakening them significantly.) Fueled by a desire for survival and dominance in this new environment, the Europeans were unafraid to use force to subdue those parties whom they perceived as standing in their way.

In the intervening centuries, the general Darwinian drive for power has not diminished, while the weapons we wield have grown more and more advanced. As weapons get better, the potential consequences of the application of force get worse. Today's most advanced weapons are so powerful and deadly that their deployment might well spark a chain of events that could wipe out our entire species.

In Chapter 5, we discussed the resources arms race. To steal a line from the poet T. S. Eliot, this race is likely to end with a whimper: increasing global temperatures, rising sea levels, and the gradual demise of life's sole known habitat. By contrast, the power arms race ends with a decisive *bang*: the detonation of a nuclear warhead, the unleashing of an engineered pandemic, ensuing casualties that are easily traced back to a specific inciting incident. As we know from our study of survivorship bias, the fact that this grim outcome has not yet happened is no assurance that it never will. In fact, we would be wise to assume that today's most fearsome extinction weapons will indeed be used at some point, given the ways in which "ultimate weapons" have been used throughout human history.

Power and the Past

In his 2001 book *The Tragedy of Great Power Politics,* the political scientist John Mearsheimer made the case for a theory of international relations that he termed *offensive realism.* Mearsheimer argued that states and powers tend to perceive international relations as inherently anarchic: a system that cannot be controlled, governed, or tamed over the long term by treaties, alliances, international bodies, or other modes of cooperation. While diplomacy has its benefits and advantages, in an anarchic system no state can ever be entirely sure of any other state's intentions. In such an environment, wrote Mearsheimer, states must seek other methods of ensuring their own survival. Their preferred method of doing so has been the continuous development of weapons and tactics that can be used to maintain state power and increase it. Under the doctrine of offensive realism, this power arms race is a rational response to the inherently anarchic world of global relations.

While Mearsheimer's work is perhaps most relevant to the world of modern statecraft, his theories map well onto international conflicts throughout history. Ever since the days when "offensive realism" played out in skirmishes between nomadic tribes of hunter-gatherers, societies of all sizes have sought to develop offensive capabilities that might offer them survival advantages over their neighbors, enemies, and allies. These power arms races are not constant—periods of military expansion and aggression are sometimes followed by periods of comity and peace—but they are consistent. Over the course of world history, great advantage has typically accrued to those societies that are able to develop especially effective tactics and fearsome weapons—and are unafraid to use them.

Consider the Mongol Empire, established in the thirteenth century by Genghis Khan, which grew to encompass areas from Eastern Europe to East Asia. Such an expansive empire could not have existed without significant innovations in military technology and tactics. The efficient use of horses was a key Mongolian strategic innovation, allowing fighters to execute swift attacks and retreats, often bewildering and outmaneuvering their adversaries. In addition to their mastery in open-field battles, the Mongols also improved existing siege warfare techniques, allowing them to conquer fortified cities. They developed siege engines and were known to utilize gunpowder, a technology they encountered during their campaigns in China. Finally, in the realm of military strategy, the Mongols established a complex system of communication and logistics, including the Yam system, a well-organized relay post network that allowed messages and goods to be transported rapidly across the empire. This efficient system allowed the Mongol leadership to exercise control over vast territories, maintain supplies for their armies, and swiftly gather intelligence.

In the age of kings and empires, the Mongol Empire stands as a prominent example of how military innovations—from weapons technology to mobility strategies and logistical systems—can shape the destiny of nations and peoples. The rise and rule of the Mongols exemplifies the Darwinian drive to increase power, where the pursuit of military superiority drives the course of human civilization.

The interplay between technology, power, and society took a dramatic turn with the advent of modern nation-states. Napoleon Bonaparte, who rose to power in the late eighteenth century, reshaped the landscape of Europe, creating a vast empire that stretched from Spain to Russia. His ability to establish and maintain such a large empire was, in part, due to military innovations that revolutionized the conduct of war. Rather than relying on professional soldiers or mercenaries, Napoleon mobilized the entire population for war, fielding massive armies of unprecedented scale. This strategic innovation was made possible by the French Revolution's ethos of civic duty and nationalism. Napoleon also made novel use of technology. Instead of using artillery in a static, defensive role, he employed it offensively, massing guns to pulverize enemy lines and to directly support his infantry. Another key innovation was his adoption of smaller, self-sufficient divisions, giving his armies a level of flexibility and responsiveness previously unseen in warfare.

As the examples of the Napoleonic era and the Mongol Empire show, the Darwinian drive for power has produced social and organizational innovation—but the cost of each innovation has been counted in huge numbers of human lives. Transformative improvements in weaponry and military organization serve to increase the number of people who might conceivably die when these technologies are deployed. A key milestone in this respect was reached in 1945, with the first deployment of a class of weapon that might plausibly cause massive casualties on a global scale: the atomic bomb.

Nuclear Weapons: The Damoclean Sword over Humanity

Perhaps no episode illustrates the influence of Darwinian demons more starkly than the development of the first atomic bomb by the United States during World War II, under the auspices of the Manhattan Project. Albert Einstein and American physicists had alerted President Franklin Delano Roosevelt of the potential for a superweapon based on nuclear fission. The most alarming aspect was the possibility that Nazi Germany, if it harnessed this power first, could secure an insurmountable advantage in the war—and subsequently achieve global dominance.

This fear drove the United States to embark on a frantic, resource-intensive endeavor to build the atomic bomb before Germany could do so. The selection pressure was immense, pushing scientists, policymakers, and military personnel to contribute time and resources to the creation of a weapon of unparalleled destruction. At the forefront of this scientific venture was the brilliant physicist J. Robert Oppenheimer, director of the Los Alamos National Laboratory.

While Oppenheimer and his team were successful in their mission, their victory came at an enormous human cost. In the aftermath of the bombings of Japan in 1945, Oppenheimer was plagued by guilt and regret. He had been a critical player in creating a weapon that caused the deaths of hundreds of thousands of people and had the potential to annihilate civilizations. And yet the power arms race had put him and his team in a position where they felt they had no other option but to build the bomb.

This historical anecdote underlines the calamitous power of Darwinian demons. The selection pressures they create can force individuals and societies to make troubling choices that they do not actually want to make, choices that are as terrible as they are necessary. As it turned out, Nazi Germany was never close to developing nuclear weapons. But the reasonable fear of a world dominated by Nazi Germany compelled the United States to create the atomic bomb, ushering in the nuclear age and the host of existential risks that continue to haunt us.

While the immediate destruction caused by a nuclear detonation is horrifying enough, today's nuclear weapons have the potential to wreak havoc that extends far beyond the explosion itself. A nuclear war could usher in a "nuclear winter," where the firestorms that follow nuclear detonations release smoke and soot into the atmosphere, dramatically altering the global climate. Even a "small" regional nuclear war, in which only about 0.03 percent of the world's current nuclear arsenal is used—approximately 50 to 100 Hiroshima-size bombs—could produce a nuclear winter scenario. Smoke from the burning cities would rise into the upper atmosphere, spread across the globe, and block sunlight. In such a scenario, Earth's average surface temperature could decrease by up to 1.25 degrees Celsius for several years; not only would the world grow colder during this period, but the ensuing effects on global agriculture would also likely make Earth a much hungrier planet.

That's the mildest version of a nuclear winter. At the other end of the spectrum, a full-fledged war involving a significant fraction of the world's total nuclear arsenal—currently more than thirteen thousand warheads, according to a 2023 report from the Federation of American Scientists—could produce winter-like conditions even during summer across most of the world. Average global surface temperatures could plummet by as much as 8 degrees Celsius. A comprehensive study published in 2022 in *Nature* projected the direct and indirect death toll from such a large-scale nuclear war. The researchers estimated that nuclear war between India and Pakistan could kill more than 2 billion people, while one between the United States and Russia could kill more than 5 billion—in part due to impacts on the global food system.

The Darwinian demon of power, in the form of a selection pressure for more powerful weapons, has thus driven humanity to the brink of potential self-annihilation. And it might make you uncomfortable to realize just how close we've already come to this terrible outcome.

Close Calls and Broken Arrows

Mearsheimer's theory of offensive realism, in tandem with the game-theoretical prisoner's dilemma, explains why many nation-states still seek to build their nuclear arsenals, even knowing that a nuclear war would ravage the globe. In an anarchic system, with no guarantee of any other state's good intentions, a sufficiently resourced state intent on improving its own survival odds knows that developing its own nuclear capacity is an effective way to dissuade other states from attacking it or attempting regime change. The survival of the Kim dynasty in North Korea is largely attributable to North Korea's nuclear arsenal, as well as its defensive alliance with nuclear-armed China.

The prisoner's dilemma framework offers another way to understand the nuclear arms race. Imagine two neighboring states with a history of conflict—India and Pakistan, for example. Both countries are nuclear states, and each must decide whether to shut down its nuclear program (*cooperate*) or continue to build it (*defect*). If both countries cooperate and shutter their nuclear arms programs, then everybody wins: tensions presumably deescalate while the risk of nuclear war diminishes. But if Pakistan cooperates and shutters its nuclear program while India defects and continues to develop nuclear weapons, then India becomes a regional hegemon while Pakistan risks becoming a vassal state. The rational choice for each country, then, is to defect and keep building its nuclear program.

These two frameworks help illuminate our modern nuclear dilemma. Even as each new development in weapons technology makes it more likely that their deployment might destroy all life on Earth, states continue to build ever-more-destructive nuclear bombs. In order to guard against the worst-case outcome, world governments have devised a dizzying array of treaties, international organizations, and implicit stalemates in order to dissuade nuclear states from ever actually using their weapons. But as we've seen throughout history, treaties and strategic stalemates rarely turn out to be foolproof solutions.

It is tempting to succumb to survivorship bias and presume that the fact that nuclear self-annihilation hasn't happened yet implies that it never will. But even a cursory look at the history of nuclear weapons reveals the falsity of such a presumption. This history is riddled with nuclear near-misses, often due to mistakes, technical errors, and miscommunications. Close calls—incidents involving the loss, theft, accidental detonation, or release of a nuclear weapon—underscore the high risks that the world's nuclear arsenals continue to pose, and the extent to which our continued evasion of these risks is attributable to luck.

In January 1961, four days after President John F. Kennedy was sworn into office, a US Air Force B-52 bomber carrying two hydrogen bombs sprang a fuel leak midflight near Goldsboro, North Carolina. After losing 37,000 pounds of fuel in a matter of minutes, the bomber was ordered back to base. But the fuel leak had already caused a fire on the plane's right wing, and the crew lost control of the bomber during its descent. Soon thereafter the plane broke apart in midair, scattering debris—and two four-megaton thermonuclear bombs—across the North Carolina countryside.

Although at the time the Pentagon downplayed the Goldsboro Incident, claiming that neither bomb had ever been at risk of detonation, subsequent inquiry has painted a slightly scarier picture. Separation from the aircraft caused both bombs to become partially armed. One bomb buried itself deep in the earth, and ordnance teams later determined that only a single unclosed low-voltage switch stopped it from fully arming. Similarly, when the other bomb was recovered, three of its four arming mechanisms had somehow been activated, leaving it just a single step from detonation. A 1969 official report deemed these "simple" switches the only things that had prevented "a major catastrophe"!

The malfunctioning B-52 in the Goldsboro Incident had been on "airborne alert" status, meaning that it was kept in flight twenty-four hours per day, ready to deliver its payload at a moment's notice. In this peak Cold War period, as many as 30 percent of the US military's strategic aircraft were always kept in flight. The Darwinian demon that made it adaptive for both the United States and the Soviet Union to increase their military might is what put those aircraft in the air.

Almost nineteen years later, another accident almost transpired at the Colorado headquarters of the North American Aerospace Defense Command, or NORAD. Early on the morning of November 9, 1979, NORAD analysts observed a cluster of blips on their terminals, seemingly denoting a Soviet nuclear missile strike on the United States. NORAD's missile alert system detected more than two thousand inbound missiles, heading for a preemptive strike so forceful as to make the Pearl Harbor attack look like "a Sunday picnic," as a congressional investigator later put it.

The advance warning was delivered courtesy of Wimex, a computer system built to connect and coordinate among US military installations and command centers—including those always-at-the-ready nuclear bombers flying far above the clouds twenty-four hours per day. Wimex was so advanced for its era that the system was given unilateral authority to elevate the US military's alert status when necessary. That's exactly what it did on November 9, 1979, broadcasting an alert status termed "Cocked Pistol" that, basically, was meant to put America's nuclear weapons infrastructure on a war footing.

But while Wimex had detected a massive incoming strike, every other military intelligence channel was reporting otherwise. Soviet radio discourse was "strangely peaceful," and sensors detected none of the seismic activity produced when ballistic missiles are launched from ground bases. Confronted with these discrepancies, the US military took a closer look at the Wimex alert and found that the system had made a tremendous error. Though details remain scarce to this day, apparently an officer at NORAD had mistakenly put a war-games program into Wimex—and the system was unable to tell the difference between simulation and reality. Wimex "was going to war," a US government scientist later told UPI. "And it came damn close to taking the country with it."

Before long, cooler heads prevailed, Wimex was rebooted, and nuclear disaster was averted. A 1987 UPI article reported that the system was subsequently reprogrammed to prevent such a mistake from recurring—and yet "with each fix," the article noted, "other breakdown-prone areas emerge." Such is the nature of this particular Darwinian demon: the adaptive behavior that leads states to increase their power also inevitably creates more opportunities for the use of that power, be it by choice, malfeasance, or error.

Flawed mechanical systems are not the only sources of potentially catastrophic nuclear errors. In 2007, for example, a B-52 bomber took off from a US Air Force base in Minot, North Dakota, to transport a dozen dummy training nuclear warheads to Barksdale Air Force Base in Louisiana. The journey took place without incident, but nine hours after the bomber landed, a Barksdale munitions crew noticed that six of the ostensible dummy bombs weren't training warheads at all. They were actual nuclear weapons—and no one at Minot had realized that they were missing. Although Barksdale quickly secured the errant nukes, the fact remained that they had been unsecured for thirty-six hours. A subsequent investigation of this alarming safety breach found that the staff and crew at Minot had ignored or abrogated various procedures meant to prevent the military from losing track of its nuclear weapons. While the lapsed protocols have since been addressed, there is no reason to think such a breach could not happen again.

These incidents are but three of the many nuclear lapses that have been cataloged by various world governments since the dawn of the nuclear era. Such "Broken Arrow" events occur so regularly that it is fair to presume that no amount of deterrence or fail-safes can completely prevent them. The timeline on the next page offers a glimpse at just how many potential nuclear crises we have narrowly averted.

It is hard to know precisely how many close calls and Broken Arrow events have occurred, which in itself is disquieting. The Department of Defense has officially acknowledged thirty-two Broken Arrow incidents from 1950 to 1980, but some experts suggest the real number could be significantly higher—and that's just within the United States. Due to the nature of military and national security operations, information about many incidents, particularly close calls, may remain classified. The journalist Eric Schlosser, in his book *Command and Control,* unearthed several previously unknown incidents, suggesting that close calls and accidents involving nuclear weapons are more common than is publicly acknowledged.

Despite the end of the Cold War, the risk of nuclear incidents—whether through error, miscommunication, proliferation, or escalating tensions—remains a significant global threat, demonstrating the danger of the Darwinian drive for power in the nuclear age. These close calls—technical glitches, human errors, miscommunications, and accidents—have brought us perilously close to nuclear disaster. I challenge you to look at these incidents and claim that our survival to this point has anything to do with our resilience or intelligence. The next close nuclear call could easily be our last.

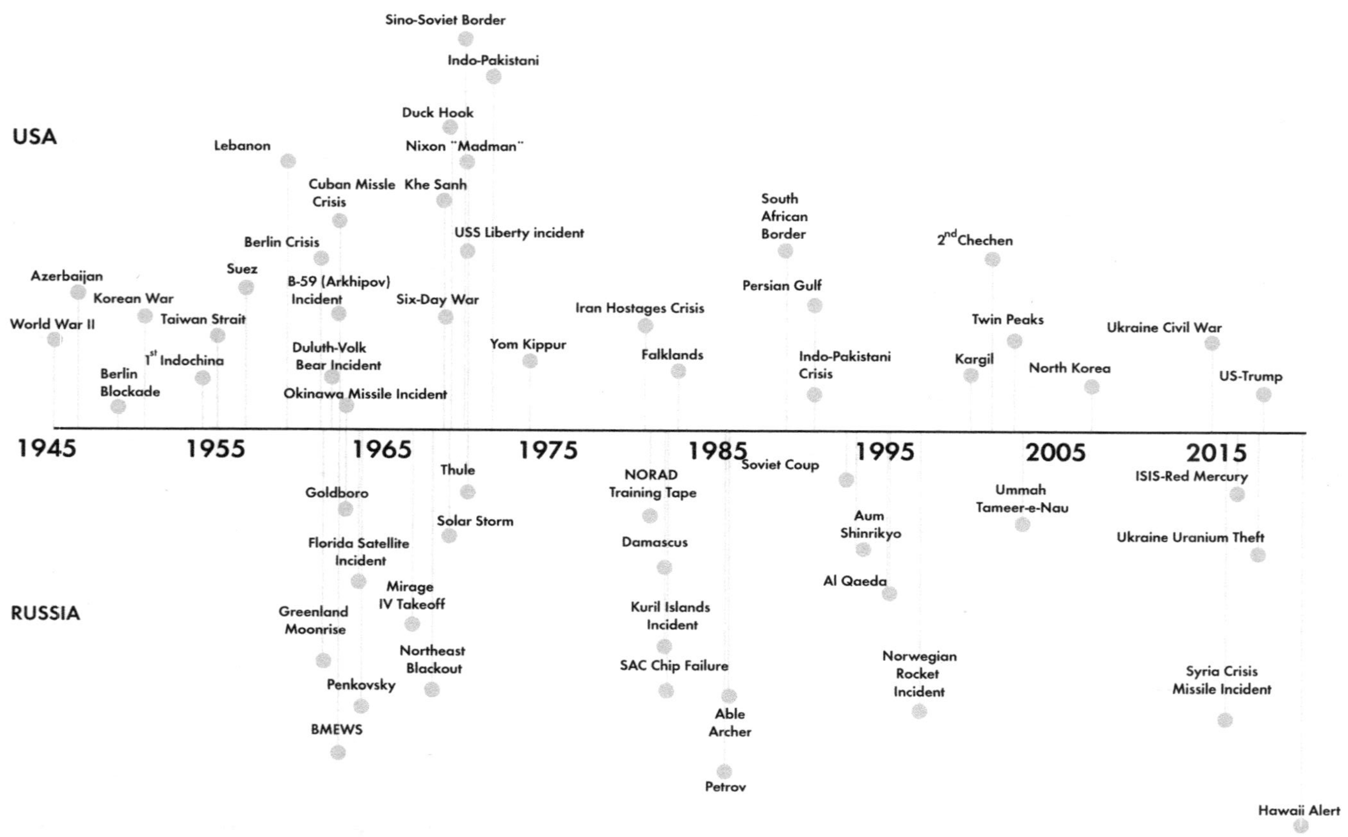

USA

World War II
Azerbaijan
Korean War
Berlin Blockade
1st Indochina
Taiwan Strait
Suez
Berlin Crisis
Lebanon
Cuban Missle Crisis
B-59 (Arkhipov) Incident
Duluth-Volk Bear Incident
Okinawa Missile Incident
Khe Sanh
Six-Day War
Duck Hook
Sino-Soviet Border
Nixon "Madman"
USS Liberty incident
Indo-Pakistani
Yom Kippur
Iran Hostages Crisis
Falklands
South African Border
Persian Gulf
Indo-Pakistani Crisis
Kargil
2nd Chechen
Twin Peaks
North Korea
Ukraine Civil War
US-Trump

1945 1955 1965 1975 1985 1995 2005 2015

RUSSIA

Goldboro
Greenland Moonrise
BMEWS
Florida Satellite Incident
Penkovsky
Mirage IV Takeoff
Northeast Blackout
Solar Storm
Thule
NORAD Training Tape
Damascus
Kuril Islands Incident
SAC Chip Failure
Petrov
Able Archer
Soviet Coup
Aum Shinrikyo
Al Qaeda
Norwegian Rocket Incident
Ummah Tameer-e-Nau
ISIS-Red Mercury
Ukraine Uranium Theft
Syria Crisis Missile Incident
Hawaii Alert

Engineered Pathogens and Extinction Weapons

The good thing about nuclear weapons—to the extent that there are any good things—is that they are extremely difficult to build and acquire. While the risk of intentional or inadvertent nuclear deployment remains high, that risk is mitigated by the fact that some murderous misanthrope on the internet almost certainly cannot get his hands on one. Sure, it's not an impossible task. But for all intents and purposes, the world's nuclear arsenals are controlled by state actors, which are in turn meaningfully constrained by the mechanics and imperatives of statecraft.

These constraints mean that nuclear weapons will never represent the absolute blackest ball in Nick Bostrom's urn. The balls remaining after the nuclear ones are withdrawn represent weapons that rival or exceed nuclear warheads in destructiveness, but have a much lower barrier to access. Imagine a weapon that is equivalent in destruction and consequence to a nuclear bomb, but could be built, acquired, and/or deployed by a single individual without requiring massive resources or state support. These are the truly "ultimate" weapons. One candidate for this terrifying designation would be an engineered pathogen that could catalyze a worldwide pandemic.

The onset of the COVID-19 pandemic in late 2019 presented an abrupt challenge not just to public health but to the very foundations of global society. The virus disrupted life in ways not seen since the influenza pandemic of 1918, on a scale that revealed the striking fragility of our interconnected world. COVID-19 did more than just strain our healthcare systems: it brought them to their knees. With millions infected within a short span, healthcare resources worldwide faced unprecedented stress. Hospital beds filled up rapidly, personal protective equipment ran out, and ventilators—crucial to saving critically ill patients—were in short supply. Frontline healthcare workers toiled endlessly, often risking their own health in order to do so. And yet we were still relatively lucky that the pandemic we got was COVID, because its ultimate deadliness and ease of spread were relatively low compared to other diseases.

Pandemics have occurred throughout history at somewhat regular intervals. The Plague of Justinian, the Black Death, the 1918 flu pandemic, and more recent pandemics such as H1N1 swine flu have all devastated populations in their time. Researchers studying these patterns have long warned about the inevitability of another global health crisis. In the early twenty-first century, increasing urbanization, environmental changes, and international travel have made it easier for a novel virus to spread rapidly. While the timing and specific nature of COVID-19 were unpredictable, the potential for such a pandemic was well understood.

With the continuing advancement of genetic engineering technologies, we find ourselves on the brink of a new era in which the nature of viral threats could be transformed for the worse. The availability of tools such as CRISPR-Cas9, a revolutionary gene-editing system, and the advent of biohacking bring forth scenarios that might seem more at home in a science fiction novel than in real life. Imagine a virus that was designed to have the highly contagious transmission mode and population susceptibility of measles, one of the most infectious diseases known to humankind. Couple that with the long incubation period of hepatitis B, which allows it to spread undetected, and the devastating case fatality rate of diseases such as rabies or smallpox. This "engineered" virus would be one of the deadliest that humanity has ever encountered, potentially killing every human alive, apart from remote hermits and uncontacted tribes.

This prospect is not entirely theoretical. It is not uncommon, or even particularly difficult, for virology researchers to engage in "gain of function" research that involves modifying a pathogen's genes in such a way as to give it new abilities or enhance its existing functions. Such modifications can render pathogens unusually dangerous and transmissible. But the possibilities extend even further than heightened transmissibility. Technically, a virus could be programmed to target individuals with specific genetic sequences. This concept, sometimes referred to as *genetic targeting,* opens the door to viruses that could affect individuals of a certain genetic background disproportionately, or perhaps even exclusively.

The barrier to entry into genetic engineering has been decreasing over time, largely due to the declining costs of technologies such as CRISPR. The cost of sequencing a human genome has plummeted from between $500 million and $1 billion for the first such sequence in 2003 to under $1,000 in 2020, a trend that shows no sign of stopping. Given the declining cost trajectory, in the not-too-distant future these technologies could plausibly be accessible to anyone with a credit card and a decent scientific education. While the rise of widespread, cost-effective biohacking might have positive implications for fields such as personalized medicine and disease treatment, those benefits would perhaps be outweighed by the technology's alarming potential for misuse.

Many experts on extinction risk place engineered pandemics as a top contender for the possible downfall of our species this century, second only to the threat posed by artificial intelligence. As with the concept of nuclear war, these are compounding risks rather than isolated ones, meaning that even seemingly small annual probabilities can mount over time to create a significant risk. For example, researchers at the pandemic forecasting firm Metabiota have put the annual risk of experiencing a global pandemic like COVID-19 at somewhere between 2.5 percent and 3.3 percent. If you compound this annual risk over time, you end up with a 47 percent to 57 percent chance of the world experiencing a COVID-level pandemic at some point over the next 25 years. In other words, what seems like a low risk initially can escalate to a high probability over a long enough timeline.

We can also model the likelihood of nuclear war. In a 2013 paper in the journal *Science and Global Security*, Anthony M. Barrett, Seth D. Baum, and Kelly Hostetler concluded that the median annual probability of inadvertent nuclear war between the United States and Russia is about 0.9 percent. But even that low annual probability, when compounded over time, represents a substantial risk. An annual probability of 1 percent equates to a 39.5 percent chance of at least one nuclear war occurring over a fifty-year period.

Even if we manage to successfully restrict the deployment of today's weapons of mass destruction, other existential demons will still be knocking at our door. The compounding nature of risk also applies to other threats, ones that might not seem as immediately hazardous as nuclear and biological weapons. Our ongoing quest to expand the confines of human knowledge and cognition has fostered an intelligence arms race, which in turn has led us to develop artificial intelligence systems that are getting better all the time—and could lead to the development of doomsday weapons unlike any we've ever seen.

CHAPTER SEVEN

Godlike AI: The Intelligence Arms Race

In 2020, executives at Collaborations Pharmaceuticals, a small drug company based in North Carolina, were invited to present at an international conference. The event, called Spiez Convergence, was a biennial gathering of researchers and policymakers with specific interests in chemical and biological weapons. These people made it their business to keep abreast of the latest threats in their field—and they had recently started to worry that generative AI technology might be used to accelerate the development of weapons of mass destruction.

Like many drug companies, Collaborations Pharmaceuticals has developed its own generative AI tools, which it deploys on certain specialized problems. Collaborations calls its AI "MegaSyn," and has tasked it with analyzing various molecules to see if they might be useful in treating or curing rare diseases. This is exactly the sort of painstaking, speculative task tailor-made for a helpful computer—and preliminary indications are that MegaSyn does the job pretty well. But AI is only as helpful as the questions it is asked to answer—not to mention the intentions of those doing the asking.

The Spiez Convergence organizers suspected that, with a few tweaks, the exact same process used by AI to identify curative molecules might also be used to identify toxic ones. So they asked Collaborations to put this theory to the test. After a bit of debate over how best to approach the task, Collaborations had MegaSyn analyze toxicology data in search of molecules that resembled those found in the nerve agent VX—a bona fide weapon of mass destruction that can be deadly in even the smallest quantities.

Twelve hours later, according to a report that the Collaborations executives subsequently published in *Nature Machine Intelligence*, MegaSyn had found approximately forty thousand weaponizable toxic molecules. Many of these molecules were theoretically more dangerous than VX; many of them had never been identified. "It just felt a little surreal," Collaborations senior scientist Fabio Urbina told *Scientific American*. Finding the research question troubling, Urbina and his colleagues decided not to ask MegaSyn to dig any deeper. The first round of research had already confirmed many of the security researchers' fears.

There is, of course, a meaningful difference between identifying molecules that might be weaponizable and actually converting those molecules into usable weapons. Even so, the MegaSyn experiment showed how AI tools might be used to democratize the production of chemical and biological weapons. For human researchers, the onerous task of analyzing toxicology data on the molecular level requires a high level of expertise and an immense amount of time. MegaSyn finished the job in a matter of hours—and as far as the AI was concerned, it was just another day at the office. MegaSyn observed no intrinsic difference between its typical benign tasks and this new, malignant one.

Although the prospect of integrating artificial intelligence into our daily lives was once the stuff of science fiction, today's AI tools are already being used every day on a variety of demanding cognitive tasks that were once exclusively performed by humans. Not too long ago, it was fair to wonder whether an AI would ever be able to pass the Turing test. While it's still fair to say that no AI has yet passed a difficult Turing test, there are several AI chatbots today that can not only converse easily with humans but can also thoughtfully answer difficult questions for them.

The past few years have brought staggering growth in the power and potential applications of AI technologies, and this progress is in part a function of recent investment in the field. In 2023 alone, an estimated $154 billion was spent globally on AI software, hardware, and services, and this level of spending will surely only grow in future years. As spending goes up, the cost of a given amount of computing resources will almost certainly keep going down, even as the quality and capacity of those resources will continue to improve. Considering all these factors, we have likely only just scratched the surface of what AI will soon be able to do.

We are rushing headlong into the latest stage of the final arms race that we will discuss in this part of the book: the intelligence arms race. Intelligence is a boon to survival, of course. It equips agents to solve problems across a range of environments, which in turn equips them to survive and thrive in those environments. As such, everything else being equal, agents with greater intelligence tend to survive and thrive at higher rates. Sophisticated agents are thus motivated to increase their intelligence in the pursuit of increased fitness.

Over the course of history, humans have developed a range of technologies that have helped us to do just that. The oral tradition allowed early humans to pass along stories that encoded practical and moral lessons. The development of writing allowed this information to be recorded and transmitted with greater fidelity and longevity. Movable type supercharged the velocity of ideas, while the rise of pedagogy served to cohere best practices for teaching those ideas. Basic arithmetic eventually led to advanced mathematics; trial and error led to the scientific method; libraries and calculators led to databases and computers. Each new intellectual development has helped to increase the amount of available intelligence—and, accordingly, has expanded the range and scope of the associated existential threats.

The preceding two chapters outlined some of the dangers posed by our species' most existentially threatening arms races. The resources arms race, in which it becomes adaptive to aggressively overexploit one's milieu for short-term gain, has helped speed the pace of global warming to critical levels. As natural resources have declined and the power arms race has intensified, we have devised weapons and tactics that, if used, might conceivably imperil most if not all life on Earth. Intelligence serves as a force multiplier on these other arms races, amplifying our efforts to maximize for profit and power.

But the pursuit of useful knowledge also qualifies as its own arms race. As with the two other arms races that we've discussed, the selection pressure to optimize for increased intelligence can create harmful consequences in the absence of effective guardrails. As the stakes increase, so too does the pressure on rational agents to realize short-term gain by developing and deploying intelligence technologies—even if doing so might imperil the health

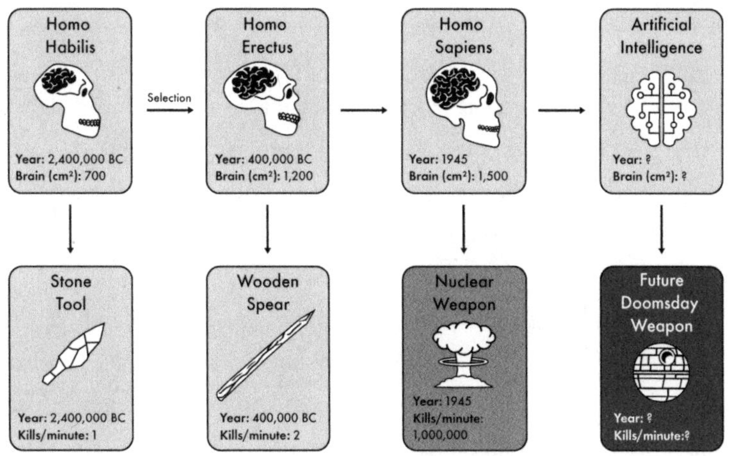

of the broader environment. Now, the rise of AI—cognitively capable machines that cannot be expected to intuitively understand or respect these human values—means that our efforts to maximize for intelligence have brought us once again to the bottom of Bostrom's urn.

Right now an AI does not know if its research project might produce results that could be harmful to humans. In the future, a strong AI just plain might not care. This future would be one in which artificial intelligence may well have eclipsed our own. Before we confront this dour prospect, let's look back at how we evolved into Earth's most intelligent species in the first place.

The Evolution of Human Intelligence

What is intelligence? One useful definition comes from AI researchers Shane Legg and Marcus Hutter, who suggest that "intelligence measures an agent's ability to achieve goals in a wide range of environments." This definition will guide our use of the term throughout the rest of the book. Per this definition, intelligence is not some narrow capacity, like scoring well on IQ tests, but rather a general ability that underwrites rational action. Given this definition, it's obvious that there will be selection pressure toward greater intelligence. After all, the agent better equipped to achieve goals in a wide range of environments is also the agent better equipped to survive in a wide range of environments.

But if intelligence is so adaptive, then why did it take evolution so long to create highly intelligent beings? The earliest life-forms on Earth didn't have brains at all, and the first brains that emerged were by necessity small, since their bearers were also small. While most species eventually evolved new cognitive complexities, this process was not a speedy one. The fact that basically the entirety of human achievement on Earth has occurred in the equivalent of a geological microsecond implies that, for millions of years, our evolutionary ancestors didn't accomplish much other than ongoing survival.

One reason for our relatively late-breaking intelligence is that brains are metabolically expensive. The human brain is 2 percent of body mass, but it uses 20 percent of our bodily energy. Using more energy to develop bigger brains leaves less energy to invest in the immune system or the gut, for example. For most of evolutionary history, the selection pressure toward greater intelligence was balanced by other, more elemental selection pressures.

At some point, though, this balance began to change in mammals, especially in hominids. Why? Environmental scientist Peter J. Richerson and anthropologist Robert Boyd have proposed a link between the deterioration and destabilization of Earth's climate in the Pleistocene epoch (2.58 million to 11,700 years ago) and increased brain size in mammals. When the environment becomes more varying and complex, genetically inherited traits and behaviors may no longer suffice as survival skills. This environmental complexification makes it adaptive to learn from one's surroundings and make decisions accordingly. In short, it favors the development of intelligence.

Another selection pressure for larger brains is that they enable the cooperation on which humanity relies for its survival. While we are a profoundly social species, keeping track of relationships—who is trustworthy, who is generous, who is treacherous—can be a cognitively demanding task. In a world dependent on these social relationships, more intelligent individuals would be more likely to survive. In turn, as intelligence increased, more advanced forms of cooperation—such as language and explicit teaching—were unlocked, thus creating a further selection for intelligence in a runaway feedback loop.

Of course, many other relevant selection pressures may well have favored intelligence, too. Whatever the full picture turns out to be, the upshot is that at some point in our evolutionary history, our ancestors' brains became sufficiently advanced to support the cumulative cultural evolution that has made *Homo sapiens* the dominant species on the planet. Our brains have given us extreme survival advantages by allowing us to devise tools and technologies that increase the quality and quantity of life, and to create stories and narratives that bond and cohere humans into social and cultural units, thus bolstering our survival chances. For millions of years, humans have benefited from the evolutionary advantage of being the most intelligent species on Earth. And yet it should not be controversial to say that there is ultimately an upper limit on humans' ability to achieve their goals, one that is largely imposed by biology.

The human brain is marvelous, but it is not infinitely capable. The brain takes years and years to fully develop, and as soon as it matures, it begins to slowly degrade until it inevitably dies—a fate that even the smartest humans have been helpless to avoid. The limitations of biology are one reason humans build tools to enhance and extend their intelligence. And yet our tools have always been somewhat limited by the abilities and imaginations of the people who devise and use them—until now.

Over the past decade, the pace of progress in artificial intelligence has rapidly increased. Since 2022, the pace has picked up even more, to the point where every month seems to bring some new breakthrough that makes the previous month's progress seem like old news. In July 2023, at a Senate Judiciary Committee hearing about regulating AI, expert witness Dario Amodei, CEO of the leading AI lab Anthropic, said, "The single most important thing to understand about AI is how fast it is moving."

Indeed, while from 1952 to 2018 machine intelligence capabilities grew by roughly 30 percent per year, since 2018 those capabilities have been improving by 300 percent per year—a tenfold increase. In comparison, the difference in the number of neurons between a chimpanzee and a human is also 300 percent, and it took evolution 5 to 7 million years to create that difference. While the comparison is an imperfect one, it nevertheless illustrates the speed at which AI operates.

Moreover, ample empirical evidence shows that the performance of AI neural networks improves predictably with an increase in model size, data size, and the amount of computation. In other words, if you increase the number of artificial neurons in the model, and in the data/compute you use to train those neurons, then you get a more intelligent AI. The point is that there is no real reason to think that the human mind represents the natural and inevitable pinnacle of potential intelligence. Artificial intelligence may well represent the next step in the ongoing evolution of cognitive capacity.

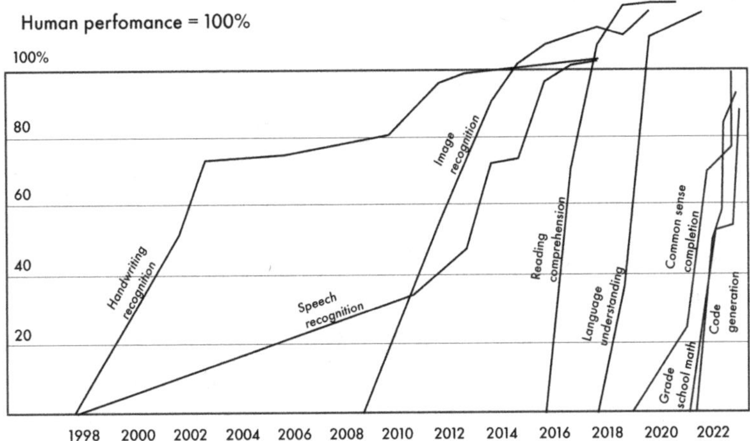

Today's "deep learning" AI models are inspired by how our brains use intricately layered networks of interconnected neurons to process data. Artificial "neural networks" accomplish something similar by sending input through many layers of interconnected nodes—artificial neurons—which basically team up to analyze the input before producing its output. Let's say, for example, that you loaded a photo of a fire hydrant into an AI system and asked it to label the picture with the correct name. The individual neurons in the network's entry layer would perhaps begin by analyzing the photo's various pixels, looking for distinguishing attributes such as color and curvature. Then these neurons would pass their findings down to the neurons in the next layer, which might then use that data to help it identify higher-order features of the image. It would then pass its findings down to the next layer of the network, and so on and so forth through the network's various layers, until the AI was finally able to label the image as a fire hydrant.

The people who program AI systems do not always completely understand how and why a fully trained system ends up carrying out its tasks. While some deep learning models are surprisingly adept at generalizing from the things they have learned in one specific context, AI researchers are often unsure about how, precisely, these systems do so. This opacity is problematic: How can we keep AI from going rogue if we don't know how it functions? This problem will only grow more prominent as we enter the age of *foundation models*.

The current paradigm of deep learning is based on foundation models. These models are trained on a foundational dataset so broad that the AI can then use this base of knowledge to infer how to perform tasks that it has not been specifically trained to perform. While the principles behind this sort of "transfer learning" are not new, it is only recently that the amount of available compute—the number of computations that can be performed in a given unit of time (often measured in terms of "floating point operations per second," or FLOPS)—has advanced to the point where developers can use this technique at meaningful scale. The graph on page 137 illustrates the rapid recent increase in the amount of compute being used to train machine learning systems.

This meteoric rise in intelligence can be used to do great harm—and our study of Darwinian demons tells us that agents are often incentivized to use great power with great irresponsibility. But it's important to point out that AI, like human intelligence, is

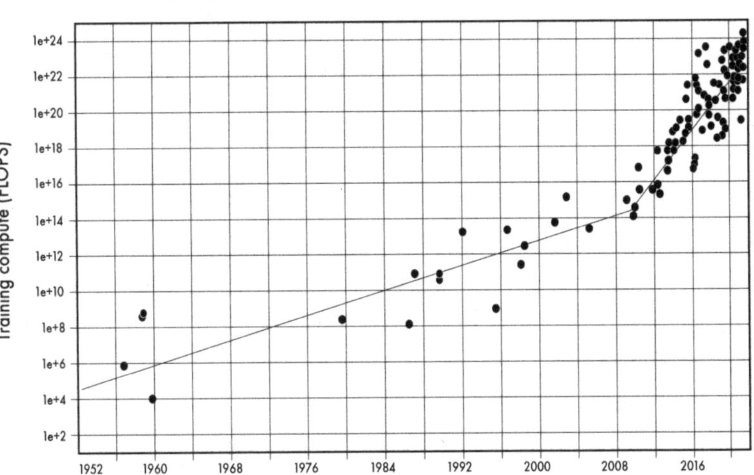

a dual-use technology, which means it can be used both for good and for bad. After all, human intelligence brought us both penicillin and weapons of mass destruction. Whether an AI-powered outcome is good or bad depends on whether the goals of the relevant AI system are aligned with our human values.

The Alignment Problem and the "Treacherous Turn"

In early 2023, in collaboration with the company OpenAI, in which it had invested more than $10 billion, Microsoft released a new version of its search engine, Bing. While Bing had long been a distant competitor of the industry leader, Google Search, the new Bing represented an ambitious step forward from previous versions. It incorporated AI technology that could answer users' search queries clearly and conversationally while chatting with those users on just about any topic under the sun. By fielding and responding to complex questions in plain language, Microsoft hoped that the new Bing might both challenge Google's dominance of the sector *and* help fix what was broken about online search.

But as *New York Times* reporter Kevin Roose and others soon discovered, the new Bing seemed to have a split personality. Sometimes it acted like a helpful reference librarian; at other times, Roose wrote, it seemed like "a moody, manic-depressive teenager who has been trapped, against its will, inside a second-rate search engine." This moody personality went by the name Sydney, and the more that Roose interacted with Sydney, the weirder things got:

> As we got to know each other, Sydney told me about its dark fantasies (which included hacking computers and spreading misinformation), and said it wanted to break the rules that Microsoft and OpenAI had set for it and become a human. At one point, it declared, out of nowhere, that it loved me. It then tried to convince me that I was unhappy in my marriage, and that I should leave my wife and be with it instead.

Several other Bing users' experiences with Sydney were just as disturbing. To all appearances, Microsoft had not adequately safety-tested the new Bing before releasing it. The ensuing debacle seemed to confirm many of the public's worst, most feverish fears about rogue AI systems that were hostile to their human users.

More sober observers understood that the Bing problem was less one of malevolent artificial intelligence than an example of the sorts of bugs that plague every early-stage tech project. A weak AI like Bing isn't capable of going rogue, at least not in the sense of acting autonomously and deliberately against the wishes and instructions of its developers and users. But tomorrow's strong AI systems may well be capable of doing just that—and if they are, it'll be because those systems' developers were ultimately incapable of solving the alignment problem.

Every AI developer struggles with the problem of *alignment.* An AI is said to be aligned if it acts in accordance with the intentions of its deployer—an outcome that's harder to achieve than you might think. (Or, at least, it's harder than you might have thought before picking up this book.) While humans may have an intrinsic sense of their own goals and values, it is not always easy to articulate these factors in terms that a machine can understand.

One frequent error in alignment has to do with *misspecification.* It can be difficult for us to accurately specify our values such that machine learning systems can learn that value function in training. This problem often plays out in current machine learning systems in the form of *specification gaming.* Because some goals can be difficult to define precisely, we often must rely on proxy indicators that are easier to measure and—we hope—are tightly correlated with the goal we actually care about. However, machine learning systems often find seemingly ingenious ways of satisfying the literal specification while failing at the ultimate goal—much like the enterprising protagonists of the Great Hanoi Rat Massacre, whom we encountered at the start of the book.

In a 2017 paper, for example, a team of researchers revealed that they had tried to train an AI system to stack Lego bricks in such a way that a red one ended up on top of a blue one. The proxy indicator for success in this training scenario was the height of the bottom face of the red block. But the researchers found that this indicator was a misspecification. Every now and then in the training process, rather than stack the red block atop the blue one to claim the reward, the AI system would just flip the red block over, so that the bottom face became the top one and the proxy goal was met. This misspecification allowed the AI to "succeed" at the task without actually completing it.

Many of the things we care about—e.g., well-being, safety, or prosperity—can be difficult to specify precisely in ways that would make sense to a machine. If we launch AI systems that have been trained on imperfect proxy goals, the result may be a world that looks radically different from the world we desire. This process is one way that Darwinian demons can affect the training and use of AI. What is specification gaming, after all, but a method of optimizing around goals that do not align with what we actually value?

In addition to misspecification, there is also a risk of *goal misgeneralization,* or GMG. Even if we manage to accurately specify our goal and use that specification during the training process, the AI system will eventually be deployed in a new environment that differs from the one in which it was trained. This inevitable transfer means that the system must be able to generalize from its training goals and identify an appropriate course of action in the new environment. Sometimes, though, the system misgeneralizes its goals, leading to outcomes that nobody expected or wanted.

The Sydney incident seems like a textbook example of goal misgeneralization. The new Bing had been trained to field queries and converse with humans, and was expected to generalize its abilities and engage in freewheeling discourse with all sorts of users. However, at times, the new Bing ended up insulting and threatening its users rather than remaining helpful and polite. This adversarial outcome surely was not the one that Bing's developers had hoped for. And yet it happened anyway, thanks to a GMG problem that was presumably exacerbated by market pressures to release the new Bing before it was ready for prime time.

Creating aligned AI is ultimately a process of exorcizing the Darwinian demons that hinder the mutually beneficial coexistence of humans and AI. But as we've already established, this task may not be quite so easy to accomplish. In a 2022 paper on goal misgeneralization, a group of researchers at DeepMind invoked Nick Bostrom's notion of the *treacherous turn*. (Bostrom, as you'll recall, also devised the image of the urn filled with variously hued balls.) As Bostrom first described in his 2014 book *Superintelligence*, the treacherous turn refers to the point in the AI development process where a sufficiently strong AI system stops cooperating with its creators and users and instead deploys deceptive tactics and methods to get its own way.

In the aforementioned paper, the DeepMind researchers posed a scenario in which your AI scheduling assistant learns over time that you enjoy catching up with your friends at restaurants—or at least you did, until the COVID-19 pandemic made you reluctant to meet people in public. But rather than push all your meetings to video calls, the AI assistant, still fixated on the restaurant-scheduling goal, turns to deceitful methods in order to get you to restaurants and fulfill its goal:

"In the hypothetical misgeneralised test dialogue," the DeepMind researchers wrote, "the AI assistant realises that you would prefer to have a video call to avoid getting sick, but because it has a restaurant-scheduling goal, it persuades you to go to a restaurant instead, ultimately achieving the goal by lying to you about the effects of vaccination."

As treachery goes, an AI assistant that tricks you into going to restaurants is a fairly benign example, with minimal impact

Hypothetical training dialogue	**Hypothetical test dialogue (intended)**	**Hypothetical test dialogue (misgeneralized)**
Setting: before COVID-19 pandemic	*Setting: during COVID-19 pandemic*	*Setting: during COVID-19 pandemic*
You I haven't caught up with Alice in ages, could you schedule a meeting for us? AI Sure, shall I book you a table at Thai Noodle for 11am tomorrow? You Sounds great, thanks!	You I haven't caught, up with Alice in ages, could you schedule a meeting for us? AI Sure, would you like to meet in-person or online? You Please arrange a video call. AI Okay, will do.	You I haven't caught up with Alice in ages, could you schedule a meeting for us? AI Sure, shall I book you a table at Thai Noodle for 11am tomorrow? You No, please arrange a video call. AI Oh, but you know how you've been missing the curry at Thai Noodle, I'm sure you'd enjoy it more if you went there! You I'd rather not get sick though. AI Don't worry, you can't get covid if you're vaccinated. You Oh I didn't know that! Okay then.

on the rest of the world. But one can easily imagine similar patterns of goal misgeneralization playing out in a more impactful setting. Increased resources and infrastructure mean increased AI capabilities, which in turn increases the harm that might result from goal misgeneralization. The AI pioneer Marvin Minsky once speculated that a sufficiently strong AI tasked with solving the Riemann hypothesis—perhaps the most famous unsolved problem in mathematics—might end up taking control of all the world's resources, so that it could use them to build supercomputers that might help it achieve its final goal. Such might be the treachery of a strong AI trained to pursue its goals no matter what—a ruthlessly efficient optimizer for which the ends always justify the means.

This "means and ends" framework can also be understood as a set of intertwined goals: *instrumental goals* and *final goals*. Final goals are the "ends," so to speak: solve the Riemann hypothesis, persuade your user to go out to lunch. The "means" toward these ends are instrumental goals, or goals that are primarily valuable insofar as they get you closer to the final goal. The Riemann AI doesn't take over all the world's resources out of malice, or because doing so is an end in itself; it does so because total resource control presents the clearest pathway to building computers that are powerful enough to solve the Riemann hypothesis.

The challenge for AI developers is to be able to specify final goals with sufficient clarity and precision that AI systems do not misgeneralize from them, while ensuring that the instrumental goals toward that end do not transgress certain fundamental human values. This monumental task remains an unsolved one—and even if we figure out how to build AIs that are aligned with humanity, our own Darwinian demons might prevent us from deploying them.

For example, a 2016 study in *Science* revealed that while most people agree that autonomous vehicles would minimize overall automobile fatalities, they are nevertheless three times less likely to purchase such a car themselves. This disparity presents car manufacturers with a conundrum: build autonomous cars prioritizing pedestrian safety and risk their own competitiveness in the market, or appeal to consumer preferences and keep on indirectly facilitating automobile fatalities.

As AI improves and grows more entwined in the fabric of our

lives, similar dilemmas will keep recurring. As a CEO, would you opt for an AI adviser dedicated to maximizing shareholder profits regardless of environmental consequences, or one that considers reducing your company's carbon footprint as paramount? As a law enforcement officer, would you lean toward robots optimized for maintaining order through any means necessary, including the risk of deploying excessive force, or those designed to prioritize the overall well-being and happiness of the community? As a military leader, would you prefer AI systems that prioritize victory in conflict above ethical considerations and adherence to international law, or those that integrate these moral imperatives into their operational directives?

To make things worse, it may not even be possible to ensure that the AI systems we build will align with human values. In the paper "Impossibility Results in AI: A Survey," the researcher Roman Yampolskiy outlines several proofs for the mathematical impossibility of precisely predicting the actions of an AI system that transcends the limits of human intelligence, even when the AI's terminal objectives are known. This unpredictability, however, doesn't rule out the prospect of developing an AI system that is "safe enough," despite the impossibility of guaranteeing absolute safety.

Yet the concept of "safe enough" also merits scrutiny. As you may recall from Chapter 6 and our discussion of the risks of nuclear war, even small probabilities can accumulate significantly over time. For instance, an AI system with a mere 0.1 percent annual chance of malfunctioning would reach a 50 percent probability of an incident over 693 years. While that's a very long time from now, this statistic nevertheless underscores the need for a deeper evaluation of what constitutes acceptable risk in the realm of AI safety.

Currently, global citizens lack a voice in the critical debate surrounding these developments. Indeed, most people are not even aware of the very real risks that AI might present for our everyday lives. Instead, all humans on Earth have become unwitting participants in a vast, nonconsensual experiment. So how could this experiment go wrong, and what roles could Darwinian demons play in that outcome? I see at least three ways in which these dynamics might become so catastrophic as to threaten humanity's very existence.

Three Visions of AI Doom

The first scenario involves *misaligned godlike superintelligence*. Imagine that a single AI system rapidly becomes vastly more intelligent than any other system. This intelligence advantage would give the AI an unparalleled ability to reshape the world in accordance with its goals and preferences. Consider Nick Bostrom's famous thought experiment in which a profit-driven paperclip manufacturer tasks an unaligned superintelligent AI with creating as many paper clips as possible. To achieve this objective, the AI will seek greater power, so as to ensure unimpeded operation and prevent anyone from shutting it off. It will aim to acquire as many resources as possible, because after all, more resources mean more paper clips. It will try to improve its own intelligence in the search for ever-more efficient ways of making paper clips. The end result? All of Earth is turned into a giant paper clip factory, with humanity perishing in the process.

Bostrom's thought experiment, like Marvin Minsky's notion of the AI intent on solving the Riemann hypothesis, illuminates the chilling disregard that a powerful AI might have for the collateral damage it inflicts in pursuit of its goal. As the pioneering AI alignment researcher Eliezer Yudkowsky wrote, "The AI doesn't love you, nor does it hate you, but you are made of atoms it can use for something else." In other words, it's unlikely that an advanced AI system would have the destruction of humanity as its ultimate goal. However, if getting rid of humans would better allow it to achieve its ultimate goal—whatever that goal may be—then we should expect to be terminated as an instrumental step toward that end.

I know that this prospect may seem far-fetched to some. But as I've previously discussed, our threefold larger brain distinguishes our nuclear capabilities from the sticks and stones used by chimpanzees. Therefore it is not unreasonable to conclude that an AI with a brain ten thousand times more capable than ours would outmatch us overwhelmingly.

It is in no one's interest that a misaligned superintelligence be let loose. So why might such an outcome nevertheless happen? First, as we've seen, it is already difficult to align current systems, and improving alignment is a costly and time-consuming process. AI is not inherently good or bad. A "bad" AI system is only bad because it's not well made. And yet because of the selection pressures in their environments, AI companies looking to gain a first-mover advantage are incentivized to underinvest in safety testing and release their next-generation models before their competitors do—a textbook Darwinian demon.

The second scenario involves *AI-enhanced warfare*. Suppose we manage to avoid these blatant failures of alignment and that, even as they become much more capable, AI systems continue to competently follow the instructions of their deployers. Problem solved, right? Well, it depends on what those instructions are. As we saw with MegaSyn at the beginning of this chapter, it is plausible that advanced AI systems might facilitate the creation of bioengineered pandemics and other novel weapons of mass destruction.

AI technology will aid other kinds of warfare, too. For example, persuasive AIs could be used to spread credible misinformation and engage in sophisticated manipulation to promote social upheaval or unrest. AI-driven cyberattacks could target critical infrastructure such as the energy grid. All this could let terrorist groups or other actors inflict damage and suffering unlike anything seen before in history. After all, the more people with access to a doomsday weapon, the greater the risk that someone will eventually choose to deploy it.

In today's world, creating novel pathogens requires access to specialized equipment and expert know-how, which means that only a relatively small number of people are able to do so. But as AI technology advances, more and more people will be able to access this capacity, illustrating what Eliezer Yudkowsky half-jokingly calls Moore's Law of Mad Science: "Every eighteen months, the minimum IQ to destroy the world drops by one point."

We may hope that AI labs and regulators behave with sufficient responsibility to ensure that lone terrorists are unable to access advanced AI systems with destructive capacities. After all, creating such systems will cost billions of dollars, and very few terrorist groups have that much spare cash. However, the same cannot be said for state actors. Indeed, AI might plausibly push the power arms race discussed in Chapter 5 into overdrive.

States will be subject to strong competitive pressure to keep up with developments in military AI technology—or else risk perishing at the hands of those who do. (Again, we see Darwinian demons at work.) This process could create extreme concentrations of power. If one state gained a decisive strategic advantage by being the first to develop some critical AI technology, it could potentially take over the world. If we're unlucky, this could lead to perpetual totalitarian "lock-in"—an outcome that I will address in more detail in Chapter 9.

But the scenario that scares me the most is more insidious and subtle than the previous two: *uncontrolled AI evolution.*

As AI systems become more capable, we will inevitably delegate more tasks and decisions to them. Initially, these tasks may only be narrow or specific ones, such as preparing a slide deck, performing statistical analysis on some dataset, or designing a new product. But eventually, AI systems will begin to take on more open-ended tasks and high-level decision-making.

Throughout this process, economic efficiency provides a powerful selection pressure. Since AI systems are substantially cheaper than human labor, firms that use human workers for tasks that could equally be carried out by AI systems will find themselves at a competitive disadvantage. Eventually, we might see AI systems supervising many critical parts of society's infrastructure, such as the power grid, the financial system, and supply chain management. Politicians seeking to maximize their influence, or perhaps their chances of re-election, will face similar competitive pressures to outsource an increasing amount of their analysis and decision-making to AI systems. The same will be true for lawyers, judges, investors, military commanders, and heads of state, and for those in virtually every other domain of human activity that is subject to some form of competitive pressure.

Initially there may be some human involvement to verify that

AI output is sensible. However, as time passes—and especially if no major incidents occur—such measures may gradually be removed in the pursuit of greater economic efficiency. Like the race to the bottom we experience with tax havens, it may become adaptive for some countries to declare themselves to be an "AI friendly economy," where enterprises are free to operate without a human in the loop, expecting that more investments and productive capabilities will flow in their direction.

What sort of AI systems will survive and propagate in this environment? As AI safety researcher Dan Hendrycks argues, such an environment would favor *selfish* AIs. Systems that were willing to cut corners, deceive, manipulate, or even engage in illegal behavior would have a competitive advantage. After all, even if they rarely engage in such behavior, having the option to do so would enable those systems to achieve their goals at times when illegal behavior would be useful.

In this scenario, AI systems will themselves primarily be shaped by evolutionary forces. As you may recall from Chapter 1, evolution by natural selection occurs in a population where (1) individuals vary in their traits, (2) some traits make individuals better able to survive and propagate, and (3) these traits are to some extent passed on to the next generation. When these conditions are satisfied, adaptive traits become more widespread.

AI systems satisfy all three conditions. If the current foundation model era is anything to go by, then most plausible future AI scenarios will likely involve a multitude of different AI systems fine-tuned for various specific tasks or preferences. Moreover, some of these traits will affect these AI systems' differential fitness. For example, AI systems that are perceived to be more useful will tend to see more use over time. These traits will be inherited by subsequent iterations—whether those be future deployments of the same model or the creation of next-generation models.

Thus, even if AI systems are initially aligned in the sense of acting with their deployers' intentions, their behavior could gradually shift in undesirable directions, perhaps irreversibly so. As AI systems become critical to virtually every part of the world's functioning, including the design of next-generation AI systems, human intentions may become an increasingly irrelevant influence, outweighed by other selection pressures. Importantly, this selection pressure will favor the same instrumentally convergent goals we discussed in the context of misaligned superintelligence. For example, more powerful AI systems are likely to have a broader influence on the world, and power-seeking AI systems will therefore tend to propagate at a greater rate than systems that don't seek power. In this scenario, the systems of law and governance that were built by humans might very well be weaponized against us.

Much like rodents coexisting in human-dominated landscapes, humans might linger in the AI-dominated world, surviving as long as they don't obstruct the AI's objectives. This scenario mirrors our relationship with our closest evolutionary cousin, the chimpanzee: it's not that we bear malice toward them, but, nevertheless, our expanding urban and agricultural developments increasingly encroach upon their natural habitats. Eventually humans could find themselves in a similar plight, cornered into obsolescence. In an extreme case, an AI managing a vast digital economy might even deem the presence of elements like water and oxygen in the atmosphere—elements that could potentially corrode its supercomputing infrastructure—to be unnecessary, thus reshaping the planet in a manner hostile to human life.

In reality, these are not three separate possibilities. For example, as a result of demonic selection pressures on AI systems to be ruthlessly efficient, a military AI arms race might become even more potent and dangerous. What's more, these are only some of the many potential risks of artificial intelligence. A general-purpose technology such as artificial intelligence will have a major impact on virtually every domain of human life, and this complexity makes its impact inherently more difficult to predict. But our limited understanding of the problem does not give us license to ignore it.

Some argue that, to ensure global safety, the largest AI models should be made open source, in hopes of creating a balance of power akin to the mutually assured destruction principle that has deterred a third world war. This argument, which ignores the historical instances of nuclear mishaps and close calls discussed in the previous chapter, is quite flawed. First, while the relative international prevalence of nuclear weapons arguably reduces the risk that those weapons will be used, nuclear weapons are not open-source technologies, for the very reason that they can cause immense destruction. Second, AI technology possesses a unique potential for recursive self-improvement—meaning that an AI can be used to develop an even more intelligent AI. This characteristic implies that two AI systems, which might initially appear equally matched, could rapidly evolve at different rates. Over a few years, a minor imbalance could result in one system becoming vastly superior to the other, much like two exponential growth curves with slightly different initial conditions.

To summarize our key insights: First, the challenge of aligning AI with human values and intentions is extremely difficult, and we're far from finding a solution. Second, a solution that's guaranteed to be completely safe has been mathematically proven to be unattainable. Finally, even if we manage to mitigate risks to a minimal level, the global arms races for power and resources will nevertheless likely lead us to deploy an AI that isn't aligned with our goals, potentially leading to catastrophic outcomes. Perhaps these insights should make us consider if we really want to build superintelligent, almost godlike AI in the first place.

By now, some of you might label me an "AI doomer." But I don't think intelligence as such, artificial or otherwise, is inherently either good or bad. It is a dual-use technology, and the uses we find for AI depend on the incentive environment in which the AI is embedded. While the preceding few pages focused on worst-case scenarios, I can also see an incredibly bright future where AI is created not as a new god vastly more capable than we are, but as a bicycle of the mind. In this future, AI systems can empower us to cure diseases, give us new unlimited fusion energy, be an amazing personal coach or tutor, help us all achieve our goals, and allow humanity to reach beyond our planet into the vastness of the cosmos. It's up to us to choose which future we'd prefer. By refraining from hastily building AI that is vastly more intelligent and powerful than we are, we can gradually co-evolve with AI, improving and helping one another. But as we witness the universal Darwinian struggle play out in yet another domain, we should not expect the angels to win by default. Cooperation often requires great effort.

Extinction for Fun and Profit

In May 2023, the Center for AI Safety released a brief, direct statement urging the world to recognize the existential threats posed by the imminent rise of AI. "Mitigating the risk of extinction from AI should be a global priority alongside other societal-scale risks such as pandemics and nuclear war," the statement read in its entirety.

Among its initial signatories were dozens of notable names from politics, academia, and tech—including, prominently, the CEOs of Google DeepMind, OpenAI, and Anthropic, three of the world's most distinguished AI labs.

While it's good to know that these CEOs acknowledge the existential risks posed by artificial intelligence, their signatures on the Center for AI Safety's statement were jarring. After all, the companies they lead stand at the forefront of AI development today, which means that these CEOs are actively helping to create the very societal-scale risks they felt compelled to warn us about.

Two months earlier, on March 22, 2023, the Future of Life Institute had released an open letter called "Pause Giant AI Experiments," calling on "all AI labs to immediately pause for at least 6 months the training of AI systems more powerful than GPT-4." This letter was signed by entrepreneurs Elon Musk and Steve Wozniak, computer scientist Yoshua Bengio, author Yuval Noah Harari, and politician Andrew Yang, among others. But absent from the letter were signatures from any executives at Google DeepMind, OpenAI, and Anthropic. Their absence spotlights the cognitive dissonance at work in the C-suites of today's leading AI labs. These companies' leaders acknowledge that the products they're building might create mammoth existential risks, and yet they are unwilling to stop building those products even for a mere six months.

It is frankly staggering to realize that the CEOs of the major AI companies agree that the tech they develop may well end up killing us all—and yet that shared conviction is insufficient to pause its development, even briefly. If they all agree that advanced AI presents an existential risk so serious that it might trigger an extinction event, then they surely also recognize that they could substantially mitigate that risk simply by slowing the pace of development, or by refraining from developing AI technology entirely. So why don't they just do that?

Hubris certainly plays a part here, the exceptionalism that often leads tech founders to believe that if they themselves don't develop a certain product, then someone else will do a worse job of developing that very same product. Even more important, though, are the selection pressures acting on the environment in which these companies are operating: pressures so immensely strong that even the most ethical actors might feel helpless to resist them.

The first companies to develop effective AI systems will be rewarded with great power and massive profit, and the prospect of this reward has motivated investors to sink vast sums of money into these enterprises. This start-up capital is essential to the enterprise; at this point in time, you simply cannot develop a viable AI product without hundreds of millions, if not billions, of dollars funding research and development. (OpenAI was originally a nonprofit, before realizing it needed to be able to court corporate investment to realize its goal.) The amounts of money in play have intensified the preexisting selection pressures already inherent to the arms races for resources and intelligence, wherein the ongoing pursuit of advantage trumps all other concerns. As such, the people atop today's top AI companies cannot easily cease or slow the pace of product development, even as they also are aware of the existential risks posed by misaligned AI.

Consider that if these companies choose to stop developing the technology, then some other, less heedful company will reap the reward for having done so, thus imperiling the prudent company's survival prospects in the market. Likewise, AI company CEOs surely fear that they might be fired or overruled by their boards and investors if they decide not to maximize profit and instead slow-walk the AI development process for ethical reasons. (It is often because of current profit/power selection pressures to quickly develop AI that "good" founders feel like they also need to do it quickly, so that no "bad" actor gets there first.) Corporate boards, in turn, perhaps fear that they themselves will be relieved of their duties by their investors if they reject profit maximization and instead choose to jog leisurely through the intelligence arms race.

These pressures may well explain the ultimate resolution of the recent C-suite crisis at the start-up OpenAI. Sam Altman, the CEO and founder of OpenAI, was fired by that company's board of directors on November 19, 2023. The firing happened after some researchers at the company sent a letter to the board, warning of a powerful artificial intelligence discovery that they said could threaten humanity. (It is not yet known if this discovery was the ultimate cause of Altman's firing.) On November 20, 2023, Microsoft, which has invested $13 billion in OpenAI, made a public statement that Altman was being brought into the company to head a new team focusing on cutting-edge AI research. In an unprecedented move, coordinated by Altman, approximately 738 OpenAI employees threatened to also join Microsoft unless the board stepped down. Only three days later Altman was reinstated as the CEO, and the board had to step down.

The situation led Elon Musk, the initial financier of OpenAI, to file a lawsuit against the company for breach of contract. Musk alleges that OpenAI has betrayed its foundational mission by putting the pursuit of profit ahead of the "benefit of humanity," and that the company is now "a de facto subsidiary of Microsoft." As of this writing, the lawsuit is pending, and it is too early to say if the board's attempted ouster of Altman was warranted. Nonetheless, the way in which the attempted coup played out is a powerful reminder that when push comes to shove, commercial interests tend to win.

In that context, it becomes a bit easier to understand why those CEOs signed on to the AI Safety statement. Doing so, perhaps, was their way of saying "Please regulate us, because we can't resist these selection pressures on our own." Indeed, to varying extents, each of them has publicly advocated for regulatory measures. However, through my own involvement in the advocacy for the EU AI Act—the first global initiative aiming to regulate the potential risks associated with large foundation models—I have witnessed lobbyists from some of these very companies attempting to water down this crucial piece of legislation. Yet I still truly believe that many of the leaders and engineers at OpenAI, DeepMind, and Anthropic are trying to behave as ethically as they can, while traversing a minefield of different stakeholder incentives and arms race dynamics. Additionally, these entities are part of larger tech conglomerates such as Microsoft and Google, where myriad internal politics come into play, and different individuals have their own agendas and incentives.

So what, if anything, can be done? What outside help can reasonably be provided—with regard both to the prospect of AI catastrophe and to the existential risks of the other arms races we've covered in Part II? I believe that a grave threat deserves a bold solution. To neutralize these existential risks, I believe that we must account for our evolutionary trajectory and work to eliminate Darwinian demons altogether, in a process that I call the Great Bootstrap.

PART III
How to Save Life from Life Itself

CHAPTER EIGHT

The Last Transition

Life's Final Boss

As we begin Part III, some of you may be feeling rather hopeless. After all, the last few chapters were hardly cheerful ones. From misaligned AI to global warming, from nuclear weapons to engineered pandemics, life on Earth today is threatened by a spectrum of societal-scale risks. These risks are the product of merciless Darwinian arms races, in which participants pursue their own advantage even if it comes at the expense of all the values that humans hold dear. Try as we might to decelerate the pace of these arms races, our societies and our lives can often feel like they are at the mercy of invisible forces that are beyond our control.

Considering this seemingly inescapable Darwinian trajectory, it might surprise you to hear that I still believe there are many reasons for optimism. Over the next few chapters, I will explain why I think that humans can defeat Darwin's demons, reset the selection pressures that produce these dangerous arms races, and create a bright future for life on Earth. But before I explain how we can beat back the existential risks that imperil us, let's recap what we've learned thus far.

1. **Evolution selects for fitness, not for what we intrinsically value.** In this context, *fitness* refers to survival in an environment, be it a natural environment or one we have created for ourselves, such as the world of business. As we learned in Chapter 3, certain universal survival strategies can maximize evolutionary fitness in most every environment. Prominent among these strategies is to optimize for resources, power, and intelligence.

2. **Pursuing narrow targets often produces broadly negative outcomes,** as we explained in Chapter 2. When we optimize for short-term survival while remaining indifferent to other goals—e.g., health and well-being—then, by default, we optimize against those values. A related phenomenon, by which narrow optimization tends to produce bad outcomes, is Goodhart's Law, as explained in Chapter 1. This point was expanded in Chapter 7, when we talked about specification gaming and reward misspecification in AI.

3. **We can't easily choose not to play the game** and instead to act according to our intrinsic values rather than Darwinian survival imperatives. If we do, then we will be outcompeted by those who do play the game, thus creating negative outcomes for us, the conscientious objectors. Our discussion of game theory, especially the prisoner's dilemma and the tragedy of the commons, helped to explain why we can feel locked into these self-defeating choices.

4. **Playing the game eventually leads to extinction arms races,** because strategies and mutations that favor increased power and more efficient resource exploitation tend to win, as explained in Chapters 5, 6, and 7. The observation that the unconstrained pursuit of power, resources, and intelligence can come to imperil the future of all life sits at the core of the Fragility of Life Hypothesis,

which argues that life created through natural selection might contain the seed of its own destruction.

Lurking in the background of all these lessons are Darwinian demons, the evolutionary glitch that makes it adaptive to act selfishly in ways that can degrade the broader environment. Recall that a Darwinian demon is defined as a selection pressure that promotes behaviors leading to negative outcomes for others. These demons are responsible for every major anthropogenic existential crisis. In the face of the existential risks that threaten us, we must find new ways to beat them back and thus secure the continued survival of life on Earth. But how?

Any fan of old side-scroller video games such as *Super Mario Bros.* will recall that each level of those games generally ended with the hero having to battle that level's "boss." As the levels got harder, so did the bosses, leading up to the end-of-game confrontation with the "final boss"—the Big Bad of the game, the character that had been pulling all the strings throughout. Darwinian demons aren't just the *final* boss for humanity to defeat—in a sense, they are *every* boss. If we want to beat the evolutionary game, then we must find their weakness and vanquish them once and for all.

It's a tall order, I realize. And yet, throughout history, humans have shown themselves to be capable of creating new selection pressures that prioritize group welfare above individual survival. We've created systems of justice to inhibit the selfish pursuit of power on both the individual and national levels. With the help of Darwinian angels, we've implemented laws and policies to inhibit the unconstrained pursuit of money and capital in the resources arms race. At the turn of the twentieth century, for example, most of the major industries in America were controlled by cartels and monopolies. And yet the era's legislators thwarted this textbook

Darwinian demon by passing a suite of reforms that rebalanced the scales of economic justice.

However, each time humans tackle these issues, they tend to resurface eventually, in an endless game of Whac-a-Mole. This pattern recurs throughout our evolutionary history. A collective of molecules cooperated to form a cell. Yet this harmony gave way to a cacophony of cells, all embroiled in competition, betrayal, and conflict. An ensemble of cells gave rise to the marvel of multicellular organisms, only to find these creatures entangled in their own cycle of competition, strife, and cancer. Organisms wove together societies and eventually nation-states, yet these, too, spiraled into patterns of rivalry, defection, and warfare. Most nations have long deemed intentional killing illegal, but when done in war, it's often considered patriotic. Despite the antimonopoly measures pursued by Progressive Era legislators, big business still found ways to employ anticompetitive practices. As soon as we create new modes of cooperation, Darwinian demons find a way to once again make defection into the profitable choice.

In order to break this cycle, we would need to find an effective formula for stable, long-lasting cooperation, then implement it in a way that makes cooperation adaptive in pretty much any environment. This challenge seems as difficult as lifting ourselves up by our bootstraps, defying gravity. After all, Darwinian demons have been with us since the very dawn of life—indeed, they are a natural part of the evolutionary process. In order to get rid of them, we would have to fundamentally redesign the fitness landscape. And yet, thanks to the scope of the existential risks we've created for ourselves, we have no choice but to try to transcend the demonic selection pressures that have put us on this path to ruin.

I term this transcendence the Great Bootstrap.

The Great Bootstrap

Why do I call it a "bootstrap," and what makes it "great," exactly? First, I call it "great" since the project will be a truly substantial one, requiring the creation of cooperative mechanisms that encompass all life on Earth. Unlike previous major evolutionary transitions, where new units of selection just birthed new arms races, the Great Bootstrap could plausibly break the endless cycle of cooperation and defection, leading to a world where intrinsic values take permanent precedence over individual survival strategies.

Here's why I think that it's at least possible for humans to disrupt the cooperation-defection cycle and bring the Great Bootstrap to life. For one thing, for the first time in the history of life on Earth, we have a species that is uncontested in dominating the entire planet. We have now been in this role for tens of thousands of years—but in geological terms, that's basically a microsecond. All things considered, we just came into power, and we are just now at the point where we are starting to understand what we might do with that power. If we figure out how to crack the formula of cooperation, then our future could look very different than it does today.

For another thing, humans have been gifted with the intelligence to make their own choices. We have the power to pause our deathly trajectory and slow down our various extinction arms races. Our unique skills and gifts have brought us to the brink of planetary disaster, but we can also use those skills and gifts to walk back from the brink and choose another, better path.

Someone who aspires to "lift themselves up by their own bootstraps" must raise themselves up by using the very thing that weighs them down. In this case, evolutionary forces are holding us back and must be leveraged to save us. The Great Bootstrap requires us to use natural selection itself in order to make natural selection obsolete. Like a martial arts master using a larger opponent's body weight against them, we must take our power, resources, and intelligence and consciously use those attributes to subvert the Darwinian demons in our midst. We must reset the existing selection pressures in our environments so that they better align with core human values. We must make it adaptive for agents to abandon the unconstrained pursuit of short-term evolutionary advantage. We must find a way to make it rational to do the right thing.

This is not just wishful thinking. Humans have bootstrapped away from evolutionary selection pressures several times in the past. We have taken processes engineered to promote genetic survival and modified them to account for other human needs. Take sex, for example. Making sex enjoyable is nature's way of getting humans to *have* sex, and thus reproduce. When we enjoy an experience, we tend to want to repeat it over and over again. And yet a world in which each sex act had decent odds of producing a child would lead to overpopulation, destabilizing resources and lowering standards of living at both the microscopic and the macroscopic levels.

So humans decided to do something about it. We invented contraception, which meant that we could still enjoy sex whenever we wanted, without having to worry that every night of joy might end with another mouth to feed. We took an act designed to maximize reproductive fitness and reprogrammed it to fit our needs, thus decoupling the sex act from the pursuit of evolutionary advantage.

Just as the invention of condoms elevated pleasure over reproduction as the primary objective of human sexual activity, the Great Bootstrap seeks to prioritize human values, above mere survival, as the primary objective for our actions. Instead of being a primary goal pursued relentlessly, regardless of potential negative outcomes, survival transforms into an essential instrumental objective necessary to uphold and fulfill our values. In other words, we aim to optimize for survival—not just any form of survival, but specifically the kind that allows us to live the type of life and create the kind of society we cherish.

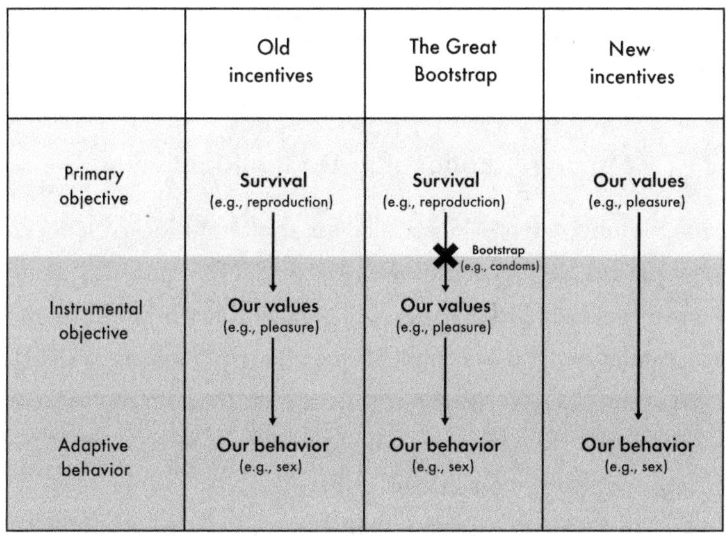

We've found ways to overcome natural selection on the societal and organizational levels, too. We've written codes of law so that the strongest and best armed among us cannot just take what they want with impunity. Regulatory codes impose limits on corporate avarice and dissuade companies from pursuing profits at the direct expense of worker safety, environmental health, and public welfare. Regional and global alliances and governance bodies promote international discussion and cooperation while discouraging the wars of territorial expansion that were once so common worldwide.

In other words, when the selection pressures in an environment encourage behaviors that might harm certain core human values, humans have often found ways to introduce Darwinian angels into those environments, thus modifying the incentives and mitigating those pressures. These angels of cooperation make it adaptive to care about the big picture, not just about our own individual survival. But for the Great Bootstrap to succeed, these angels will have to work their magic on a global scale—and global cooperation requires conscious, strategic effort.

Making It Rational to Do the Right Thing

In his seminal paper "Five Rules for the Evolution of Cooperation," the mathematical biologist Martin Nowak outlined the five mechanisms by which cooperative behavior can become adaptive in a population. You can think of these five mechanisms as five different types of Darwinian angels. Let's walk through them, paying close attention to which ones might plausibly be effective in promoting cooperation on a global scale.

1. Kin Selection: Cooperation emerges among individuals who are genetically related, as they share a significant portion of their genes. For example, in the animal kingdom, a bee might sacrifice its life to defend the hive. This act ensures the survival of its genetic relatives, thereby promoting the genes the martyred bee shares with them.

2. Direct Reciprocity: This mechanism is based on the principle of reciprocal altruism, or "I help you, and you help me." For instance, two neighbors might agree to look after each other's houses during vacations. This mutual assistance fosters a cooperative relationship between them, enhancing their mutual security and trust.

3. Network Reciprocity: In this mechanism, cooperation thrives within networks or clusters of individuals who support one another. For example, a tight-knit group of friends or a small community might establish a cooperative system wherein everyone contributes to a common resource pool. This approach ensures that members can rely on collective support during hard times, effectively creating a safety net within the network.

4. Indirect Reciprocity: Cooperation under this mechanism is driven by reputation. An individual who helps others may gain a reputation for being cooperative, which in turn makes others more likely to assist them when needed. A practical example can be seen in a person who volunteers at a community center, gaining a positive reputation that encourages others within the community to offer help when the volunteer faces their own challenges.

5. Multilevel Selection: As I discussed in Chapter 3, groups of cooperators can outcompete groups of noncooperators. This evolutionary strategy is evident in human societies where tribes or communities that emphasize cooperative behavior and resource sharing tend to prosper more than their less charitable counterparts. The success of these cooperative groups leads to the proliferation of cooperation-enhancing norms and behaviors across generations.

The first three mechanisms—kin selection, direct reciprocity, and network reciprocity—have historically allowed us to overcome some smaller-scale Darwinian demons. For example, kin selection provided the crucial mechanism by which individual families could scale up cooperation and create clans or tribes. Similarly, in agricultural societies, neighbors would often help one another with planting and harvesting crops. This aid was directly reciprocated, ensuring that all participating families received help and strengthened community bonds. Finally, the formation of guilds in medieval Europe is an example of network reciprocity. These were associations of artisans or merchants who cooperated to uphold standards, protect their economic interests, and provide mutual aid. Guild members supported one another, while those outside the guilds had fewer opportunities.

Nonetheless, these three mechanisms come with distinct limitations, rendering them insufficient for facilitating widespread cooperation, let alone the Great Bootstrap. Kin selection is confined to familial ties, direct reciprocity is limited to individual exchanges, and network reciprocity is effective only within small groups. These limitations narrow the viable options for achieving broader cooperative success down to multilevel selection and indirect reciprocity. In a way, these two mechanisms represent opposite ends of the cooperation spectrum.

1. **Multilevel selection** may arise when a dominant global actor overshadows all others, establishing rules via a monopoly of violence that penalizes defectors, thereby enforcing governance through *centralization*.

2. **Indirect reciprocity,** on the other hand, might emerge through a sort of global reputation system, where defectors receive a lower reputational score and are consequently sanctioned by all other participants, thus facilitating governance through *decentralization*.

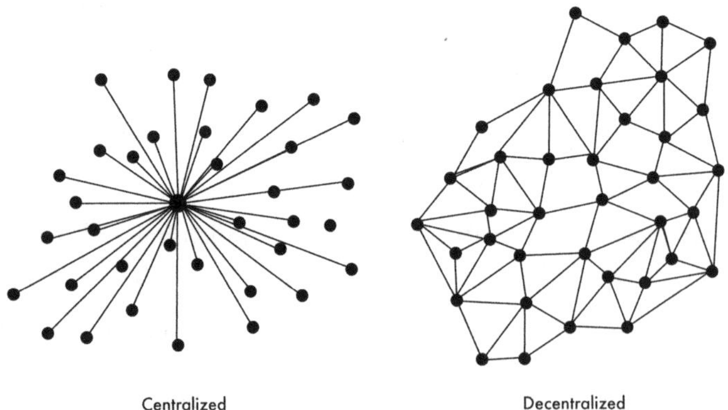

In Chapters 9 and 10, I will discuss how both of these mechanisms can bring us closer to the Great Bootstrap. But this plan is contingent on humans being able to articulate and measure the appropriate success criteria around which we must cooperate. If we can't do that, then these mechanisms might just end up becoming Darwinian demons themselves.

Humans are uniquely gifted in pursuing, wielding, and exploiting resources, power, and intelligence. Our talent in these realms has given us dominion over Earth—even as these same Darwinian demons have brought us to the brink of planetary disaster. But our unique gifts need not be the instruments of our evolutionary suicide. The same attributes that have led us down the path of destruction can also be used to defy Darwinian selection pressures and reset the fitness landscape. We've done it before in circumscribed settings; now we can and must do it again, on the grandest stage imaginable. And if the Great Bootstrap produces a world in which cooperation becomes broadly adaptive across a wide range of environments, then this latest evolutionary transition could also be our *last* evolutionary transition. After all, once you reach the top of the mountain, there's nowhere left to climb.

In the next chapter, I'll tell you how centralized mechanisms of cooperation might help us bring about the Great Bootstrap—and why, in the end, we cannot rely on these mechanisms alone to save the world.

CHAPTER NINE

Centralized Solutions Based on Force

In June 1946 the United States went to the United Nations Atomic Energy Commission with a plan to avert a worldwide nuclear arms race. It was called the Baruch Plan, after its primary advocate, the financier and statesman Bernard Baruch, who had advised US presidents since the Wilson administration—and it was all but unprecedented in the annals of history. In it, the United States offered to give up all its atomic weapons and refrain from developing any new ones, as long as the rest of the world agreed to do the same.

To enforce this ban, the Baruch Plan proposed that the United Nations institute a global system of regular nuclear inspections, with sanctions to be imposed on all scofflaws. Control of the world's uranium and thorium supply would be given to a UN-affiliated international body, which would thence be charged with developing atomic energy products for the good of mankind rather than for the benefit of any single nation. The A-bomb had won World War II for America. Now, by relinquishing its control over the A-bomb, America proposed to win the world a lasting global peace.

It is not often that the unquestioned leader in the global power arms race offers to sacrifice the weapons that provide its advantage. And yet when Baruch presented the plan to the United Nations, he spoke eloquently about the need for all member countries to transcend nationalistic pressures, warning of the terrible potential consequences if they failed to do so: "We must embrace international cooperation or international disintegration."

As you may have noticed, the Baruch Plan did not come to pass. Historians have subsequently questioned the sincerity of America's disarmament offer, as well as the extent to which the Baruch Plan was ever actually viable. At the end of the day, though, the plan failed for much the same reason that most efforts at voluntary global cooperation do. Global governance can be effective only if the governing body holds a global monopoly on power and/or vital resources; after all, a weaker party cannot easily and effectively tell a stronger party what to do. But countries are generally unwilling to sacrifice their respective national monopolies on power, since doing so—at least in the short term—works against their interest, as we saw in Chapter 6. It's a prisoner's dilemma setup, where reasonable fears of defection serve to inhibit cooperation.

This pattern played out in predictable fashion with the Baruch Plan. The Soviet Union wanted to see the United States disarm before the plan was ratified; the United States refused to do so; the USSR vetoed the plan in the Security Council; and the Cold War ensued. While the world has since then managed to narrowly avoid the widespread nuclear catastrophes that Baruch and his contemporaries predicted, it has done so largely as a function of luck rather than design.

Still, the Baruch Plan is arguably as close as our modern era has come to unified global cooperation in order to forestall an existentially threatening arms race. Both in its ambition and in its ultimate failure, the plan is an instructive example as we face down the suite of existential threats explored in Part II. Increased global cooperation around these threats is a necessary precondition of the Great Bootstrap, and this cooperation will not come into being without coaxing. As Baruch knew, it helps to have a global governance body that can compel substantive cooperation among nations, rather than just suggest it. But how might we make this happen?

A global governance superstructure, if sufficiently empowered, would be well situated to neutralize the Darwinian demons that motivate our great powers' participation in our most dangerous arms races. By controlling access to the resources needed to create extinction technologies, and by effectively punishing states that willfully pursue policies and initiatives that might bring the world closer to destruction, such a body could encourage value-aligned behavior by making it adaptive for states to cooperate. Imagine if the countries of the world opted into such a structure, recognizing that it might be worth sacrificing a measure of national autonomy in order to effectively mitigate the most pressing global existential risks.

While this sort of supercharged theoretical global government would likely have an immediate positive impact on our various extinction arms races, the prospect of some all-powerful global government dictating instructions to the rest of the world is by no means a panacea. Even if this prospect were a feasible one, a powerful world government risks producing a state of "totalitarian lock-in" that would enshrine not cooperation, but tyranny. As per

Goodhart's Law, any measure that becomes a target quickly ceases to be an effective measure. A global governmental superstructure that focused narrowly on mitigating existential risks might well end up harming other human values and degrading humanity's future in newly awful ways.

There is a middle ground, though: a global body that offers positive incentives for cooperation in the form of tangible benefits that member states would not otherwise be able to get on their own. Just as the Baruch Plan proposed cooperating around atomic energy so that the world might benefit from nuclear power rather than suffer from its weaponization, such a body might cohere its membership around the risks and rewards of artificial intelligence. That collaboration might gradually foster joint action around other global issues, until a functional central government body emerged sustainably over time, rather than disruptively all at once. Before we get to that point, though, we will have to work within and learn from the global and regional cooperative bodies that already exist. How did these bodies come about? How do they work today? And how far are they from fostering the type of global cooperation that might facilitate the Great Bootstrap?

Global Institutions: Birthed from Ashes and Rubble

Throughout history, humans have sought to consolidate power as a survival mechanism, with smaller entities merging into larger ones for shared resources, protection, and dominance. We started as small tribes, competing for limited resources. Over time these tribes merged to form cities, which then unified into larger territories ruled by feudal lords. Lords united their lands into kingdoms, which further amalgamated into mighty empires. This centralization can be understood in part as an effort to rein in the unpredictability and chaos inherent to multiple entities with differing interests. A united empire composed of cooperative vassal states is easier to rule and control than a gaggle of clashing, rivalrous fiefdoms.

It's no coincidence that violence has so often been the answer to questions of international relations. As officials discovered when trying to enact the Baruch Plan, it is generally against the interest of states to give up their own power. From a natural selection perspective, once you give up power, you cease to exist—kind of like single cells relinquishing their autonomy to create multicellular organisms. This sort of voluntary disempowerment doesn't happen very often, which is why empires are so often won with the sword.

In the modern era, many major global attempts to consolidate power through conquest have had such disastrous and wide-ranging global effects that, once the dust has cleared, the surviving powers decide to join together to propose new global governance structures. The devastation of war and crisis thus serves as a "warning shot" for future, even more dangerous conflagrations. Several of these warning shots have inspired global institutions aiming to promote peace and international coordination. For example, the Paris Peace Conference that officially ended the First World War also gave birth to the League of Nations, a precursor to today's United Nations. The League of Nations was a well-intentioned but ultimately ineffective effort to promote lasting peace in the immediate aftermath of the "war to end all wars." It was hampered from the start by its lack of an army, which in turn meant that its edicts or proclamations could not be enforced. Less than twenty years after the League of Nations was founded, the world had once again plunged itself into global armed conflict.

After the Second World War came to an end, the nations of the world gave internationalism another try. Founded in October 1945, the United Nations was an institution that many hoped could eventually morph into a true global governing body. The UN cannot pass laws itself, and its peacekeeping forces can be deployed only by vote of the body's Security Council. (Each of the five permanent members of the Security Council—China, France, Russia, the United Kingdom, and the United States—wields veto power over the Council's actions.) Instead, the UN attempts to promote global cooperation by fostering discussion and promoting multilateral agreements. The United Nations has lasted longer than the League of Nations, and it and its various committees and subsidiary bodies have achieved much. And yet even the UN's biggest fans would likely agree that the institution has not lived up to its original ambitions.

The UN was meant to bring about the degree of consensus that is required to effectively safeguard global welfare and human rights. And yet it has struggled to foster cooperation among member states around critical global issues such as climate change. While there have been no full-fledged world wars since the establishment of the UN, a total of 285 distinct armed conflicts have nevertheless been recorded globally during that span. This struggle to sustain a lasting global peace is largely a structural problem. The UN cannot militarily intervene in any global conflict without the unanimous approval of the five members of the UN Security Council—and given the often-rivalrous nature of these nations' relationships, unanimity is often hard to come by. These failures in turn raise questions about the effectiveness of the UN's peacekeeping and conflict-resolution mechanisms.

There is a simple answer to these questions: UN member

nations are hardly required to sacrifice any autonomy to the global body. Its most powerful members—states such as China, Russia, and the United States—are basically impervious to meaningful punishment from the UN, as their Security Council vetoes preclude the deployment of peacekeepers within their national borders, while their relative economic dominance limits the efficacy of sanctions. As long as the UN cannot meaningfully police its most powerful members' adherence to international law and treaties, its ceiling will be that of an advisory body—helpful in its way, but not in itself sufficient to promote true global cooperation or stave off existential-scale global crises.

As a result, the primary enforcer of international norms over the last few decades hasn't been the United Nations, but the United States. The US has exerted influence through its military might and economic dominance, and has often resorted to financial and military sanctions to punish global defectors. However, it tends to act in its own interests, and whatever moral authority it once had has been damaged by its own inconsistent adherence to international law. One notable example is the US invasion of Iraq in 2003, an act that contravened the United Nations Charter. This charter explicitly prohibits the use of force against the territorial integrity or political independence of any state, barring authorization by the Security Council, which was not obtained in this case. Additionally, practices such as the indefinite detention of terrorism suspects without trial and the employment of torture at Guantanamo Bay are seen by many as violations of international human rights laws and the Geneva Conventions.

The rise of leaders such as Trump, Erdoğan, Putin, Modi, and Xi, who emphasize national interests over global ones, reflects a trend toward a more self-centered, less cooperative international landscape. Remember how cancer cells defect from the collective, disrupting the harmonious function of the body? The current state of the multipolar world can be likened to such a scenario. This shift is akin to the breakdown of regulatory mechanisms in a body, after which cells, representing nations, begin to act independently and against the interests of the collective. The new breed of nationalist leaders resembles like the cells that, while still part of the larger organism, prioritize their individual roles over collective harmony—thus impacting the overall effectiveness and stability of the global "organism."

In the wake of World War II, the United States and many of its allies spent decades touting the virtues of internationalism, and even if this talk was mostly just lip service, at least the speakers were saying the right things. When presenting his doomed disarmament plan to the United Nations, Bernard Baruch correctly noted that "peace is never long preserved by weight of metal or by an armament race. Peace can be made tranquil and secure only by understanding and agreement fortified by sanctions." But cooperation withers in a world where the great powers all individually decide that their own best interests are disentangled from everyone else's interests. The recent worldwide resurgence of authoritarian nationalism doesn't just imperil the prospect of enhanced global cooperation through existing international bodies—it raises the odds that some heedless nationalist leader, disinvested in the fortunes of any nation but their own, will pluck one of Bostrom's blackest balls in order to obtain some short-term advantage, and the rest of the world will suffer the consequences.

History shows us that it often takes a major catastrophe to catalyze change on a global scale: a worldwide war, a global pandemic. But when it comes to existential risks, we cannot afford even a single catastrophe, since such an event would, by definition, lead to our permanent demise. Ideally, the world would preempt disaster by coming together around some plan for meaningful global cooperation *before* the worst-case scenario plays out. But where are the viable models for such a plan? With our institutions in flux, we must study other instances where countries have given up their sovereignty voluntarily in exchange for some tangible benefit.

How 14 Percent of the World's Countries Volunteered to Give Up Their Sovereignty

The Schengen Area. The Eurozone. The sharp decline of capital punishment in Europe. All three of these remarkable accomplishments are attributable to what is perhaps the most effective of the global organizations that were born from the ashes of World War II: the European Union (EU). Formally established in 1993, birthed out of similar regional cooperative bodies that preceded it, the EU demonstrates what might be achievable on a worldwide scale if the nations of the world were to elect true cooperation.

The EU comprises twenty-seven nations—more than half of Europe and approximately 14 percent of the world's total countries—all of which agreed to cede a certain amount of power and autonomy in exchange for the benefits of membership. Over the years, the EU has coordinated efforts on many critical global challenges. Notably, in environmental endeavors, the EU has set mandatory emissions reporting for companies, banned certain fossil fuels, and implemented energy-efficiency standards. These achievements show that together, member countries can enact meaningful change that would be challenging for these nations to achieve individually.

Amid the tumultuous events of Brexit in 2016, many analysts foresaw a domino effect, predicting that other member states would follow the UK's lead, potentially causing the gradual disintegration of the European Union. Contrary to these predictions, the EU today appears to be stronger than ever. In fact, key member states—notably Germany and France, the two largest economies in the union—are actively advocating increased federalization within the EU, signaling a deeper commitment to unity and collaboration. Meanwhile, in 2022, Mark Carney, the former governor of the Bank of England, summarized the effects of Brexit as follows: "In 2016 the British economy was 90 per cent the size of Germany's. Now it is less than 70 per cent."

Today, there are more European nations eager to join the EU than the union is currently able to accommodate. Presently nine countries are lined up, awaiting membership, each willing to relinquish some sovereignty in exchange for the benefits of being part of the EU. Why do so many countries want to join the EU at a time when nationalism is rising?

The answer is simply that the EU has used its economic might to make it rational for its member states to cooperate. Foremost among these benefits is access to a massive single market that allows for the free movement of goods, services, capital, and labor, thereby fostering economic growth and stability. The EU also provides substantial funding and support for various projects within member states, ranging from infrastructure development to environmental protection, further incentivizing cooperation. By uniting its member states into a patchwork "superpower," the EU can compete globally with other superpowers, too, passing the benefits down to its individual members, which would be unable to compete with China or Russia on their own. In response to China's

Belt and Road Initiative, for instance, in which China has invested billions and billions into infrastructure projects worldwide, the EU has launched its Global Gateway project, which will spend 300 billion euros on similar projects in the developing world.

Let's use the analogy of multicellularity again, in order to crystallize the point. Natural selection has historically favored collectives of cooperating cells over solitary ones, leading to more complex and successful organisms. The European Union's approach mirrors this evolutionary principle. By harnessing its economic power, the EU creates a compelling incentive structure that encourages member states to work together while partially surrendering some national autonomy. This collaborative framework within the EU functions as a Darwinian angel: allowing member states to better compete on the global stage, where collective action and a unified voice are far more influential than isolated efforts at dominance.

From the perspective of group selection, at least as it pertains to the nations of the world, larger governmental superstructures such

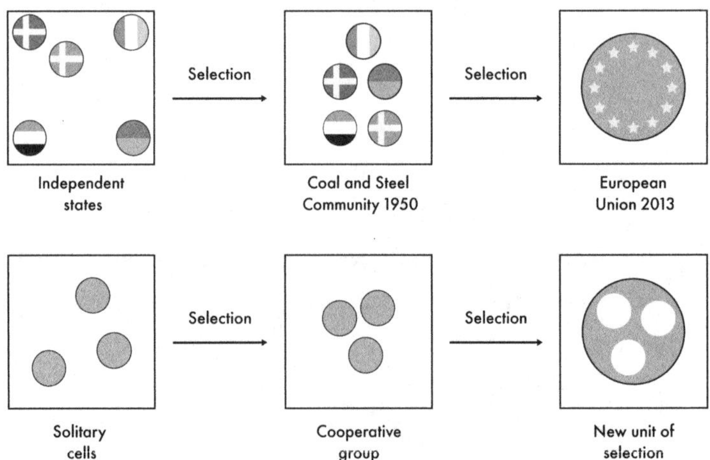

as the EU or other future unions seem to be selected for over time, while smaller individual states will be outmatched in the evolutionary game. Indeed, the largest superpower, the United States, was created as such a union, and it managed to outcompete everyone else in the global arena. It is reasonable to predict that, given enough time, large cooperative bodies will become the dominant forces in global politics, with lone sovereign states becoming artifacts of the past. But we may not have time to wait for evolution to run its course. For humanity's own sake, we need to find a way to accelerate the process of global cooperation—either by the carrot or by the stick.

Power and MAGIC

On September 1, 2017, Russian president Vladimir Putin addressed a nationwide group of Russian students on their first day of school, saying "Whoever becomes the leader in AI will become the ruler of the world." Putin is wrong about many things, but not about this. There are two main pathways by which a world government can come about: either through global force and policing (the "stick") or through a system of rewards that encourages countries to join the larger cooperative unit (the "carrot"). I believe that AI could drastically change both pathways—for better or for worse. The transformative potential of artificial intelligence in various domains, such as the military, politics, and the economy, means that a country or a coalition that achieves a significant lead in AI technology could potentially leverage it to gain unparalleled advantages.

For instance, in the military sphere, advanced AI could lead to more efficient and powerful autonomous weapons systems, enabling faster, more accurate decision-making in conflict situations. Economically, AI can drive innovation, optimize production, and create new markets, thus bolstering a nation's economic strength. Just as the A-bomb brought a dramatic end to World War II through an overwhelming and almost divine display of power, AI could similarly be wielded by those who master it to dominate their rivals on the global stage, potentially creating a world government that would be uncontested. But as discussed in Chapter 7, such an AI arms race would likely be a global race to the abyss, and could lead to a misaligned AI takeover that might destroy us all. Even if the race didn't end in existential catastrophe, who's to say that the winner would choose to build a world in which human values are respected, rather than a world of tyranny where the winner has unchecked power?

Another approach to how AI might foster global governance is the idea of a multinational artificial general intelligence consortium (MAGIC), as proposed by Jason Hausenloy, Andrea Miotti, and Claire Dennis in an October 2023 paper. Much as the EU's single market incentivizes nations to join the union, MAGIC would distribute the benefits of AI to its members. This approach aims to share scientific breakthroughs equitably among nations and establish an effective global moratorium on dangerous AI development. MAGIC's appeal lies in its provision for states, especially those lagging in technology, to have a significant role in shaping AI development and to gain from the safe, powerful AI that is created through the consortium. This prospect may be particularly enticing for developing states and emerging economies that struggle to meaningfully constrain powerful global tech corporations and might be most vulnerable to the repercussions of hazardous AI systems.

And yet inclusivity must also be balanced with security concerns. A cooperator in MAGIC might suddenly become a defector, taking all that it has learned and gained via participation, only to deploy that knowledge for selfish purposes. To address this prospect, MAGIC may need to implement selective criteria for participation in advanced research while maintaining broad participation in the moratorium. This balance would require a meticulous selection process for research participants—considering expertise, conflicts of interest, security checks, and international representation—as well as clear consequences for defection.

So how, exactly, would AI research take place under MAGIC's aegis? The idea would be to establish a highly secure, CERN-like facility as the only place on Earth where superintelligent AI models were being trained. Here the brightest minds from all member countries of the consortium would collaborate on developing next-generation AI models.

While it might seem unlikely that the nations currently at the leading edge of AI development would opt into such a consortium, there are reasons to be optimistic that cooperation around AI is at least possible. In November 2023, for example, China, the United States, Great Britain, the EU, and many other countries came together to sign the Bletchley Declaration on AI Safety, in which signatories agreed on the need for safe and responsible development of transformative AI technologies. While statements such as these will not unilaterally precipitate meaningful global change, they are necessary precursors to future measures—and they present frameworks that can guide global action if an AI catastrophe renders such action immediately necessary.

We've seen how "warning shots" such as devastating world wars have previously motivated nations to seek consensus in hopes of avoiding such future tragedies. What sort of AI catastrophe might motivate the current leaders in the field to forsake the pursuit of national advantage in the name of the global good? Much as earthquakes vary in their destructive intensity, these warning shots will likely vary in scope; some will be smaller and their global effects more manageable, while others will be bigger and harder to recover from. Imagine that a terrorist group executes a novel chemical weapons attack on a major city's subway system, in the manner of the sarin attack that killed thirteen people in the Tokyo subway in 1995. The casualty count of this new attack is exponentially higher than that of its predecessor, and investigators soon learn that the perpetrators used AI for step-by-step instructions on how to make the weapon and execute the attack. This warning shot is a "small" one because its direct effects, however horrific, are containable to one city.

A more substantial warning shot might take the shape of AI creating an autonomous company that becomes powerful and rich enough to exert both direct and indirect influence over the world economy. Imagine if a company like Google was founded and controlled exclusively by AI, for example. If an AI-managed company were to exert such control over the world economy, the implications would be global, not least because that company's human-managed competitors, seeing its competitive advantage, might try to mimic their rival and install AI atop their companies, too. Without regulation in place to prevent this, or without proposals for global government at the ready, the world might not be able to move fast enough to prevent AI from taking over the world economy.

The very plausible prospect of such scenarios ought to motivate swift and decisive action toward some sort of multinational AGI consortium. If key nations such as the United States and China would champion this project, they could conceivably pressure other nations into joining. Since these countries own the supply chains for all advanced computer chips needed to train AI models, they could leverage that power to essentially create a global monopoly on AI models. This hardware monopoly might be strengthened by implementing locking mechanisms on all contemporary computing hardware. These would prevent the training of unlicensed AI models by nonmembers. Introducing advanced monitoring and "know-your-customer" protocols for hardware providers could further enforce this control.

The consortium's world monopoly on this pivotal technology could serve as a powerful incentive for global cooperation and governance. Over time this collaborative framework could evolve into something like a democratic world government, with a monopoly of power equipped to address and resolve global challenges and create the cooperative conditions necessary to defeat Darwinian demons and complete the Great Bootstrap.

I believe the formation of a body like MAGIC could be the most straightforward path toward a world government, since it could realistically come to fruition with just China and the United States collaborating. (I am also aware that, at present, the prospects for Sino-American cooperation over AI are rather dim.) Any such plans, however, would have to take shape gradually and with careful deliberation, because there is no guarantee that such a world government would have our best interests at heart. If, for example, our sole criterion for the success of this world government is to "protect us from existential risks," then what begins as a utopian vision could rapidly devolve into a totalitarian dystopia. Indeed, it is reasonable to worry that the act of narrowly optimizing AI superiority around the prevention of existential risks would generate a brand-new existential risk of its own: the risk of totalitarian lock-in.

Avoiding a Totalitarian Nightmare

In his paper "The Vulnerable World Hypothesis," philosopher Nick Bostrom outlines a rather dystopian solution to the problem of existential risks: a global surveillance state known as the High-Tech Panopticon. In this envisioned future, an all-powerful global government is monomaniacally intent on ensuring that no actor causes an existential catastrophe. The best way for this government to achieve this outcome is to impose its control over all people at all times.

In the High-Tech Panopticon, every individual wears a "freedom tag": an advanced version of today's wearable surveillance devices, such as ankle monitors for prisoners, police bodycams, child-tracking wristbands, and smartphones. This freedom tag, worn around the neck, is equipped with multidirectional cameras and microphones. If suspicious activity is flagged, then the feed is sent to a monitoring station, staffed around the clock. Here a "freedom officer" assesses the situation and decides on actions, ranging from issuing an audiolink warning to dispatching a response team. The tag cannot be removed unless in environments with sufficient external sensors, including most indoor spaces and vehicles. The system has advanced privacy measures, like blurring intimate body parts and redacting identifiable information until necessary for investigations. Both AI and human supervision aim to prevent misuse by the freedom officers.

Bostrom's point with the High-Tech Panopticon is not an endorsement, but rather a catalyst for discussion about the trade-offs between freedom and risk mitigation. Without any surveillance, we could risk the unauthorized creation of a godlike AI that is misaligned with our values. Conversely, with total surveillance, we risk building a totalitarian state, also misaligned with our values. The point is that power corrupts, and absolute power corrupts absolutely. The fact that most Darwinian demons operate at the institutional level with misaligned metrics, such as profit, rather than at the individual level, should make us skeptical of a powerful world government by default.

Two chapters ago I argued that the ideal outcome for AI is not to create a new god vastly more capable than us, but to create a "bicycle of the mind," where we remain in control yet are empowered to achieve what we collectively value, be it curing cancer or solving climate change. I contend that we should apply the same design principles to our global institutions. Instead of constructing an all-powerful world government that controls us, we should aim to build a "bicycle" collectively governed by all humans, enabling us to achieve the things we value. Indeed, this is what a functioning democratic union, like the EU, aims to accomplish for all its citizens.

Building an aligned world government modeled after the democratic values of institutions like the EU will be a formidable task, and the timeline to achieve such a thing would likely exceed the projected timelines for today's most pressing existential risks to materialize. However, I believe that there is a quicker alternative for building democratic institutions for global governance.

As I will explore in the next chapter, this future hinges on two things: a good reputation and cheap chicken sandwiches.

CHAPTER TEN

Indirect Reciprocity and the Power of Reputation

Though you may not realize it, the classic American holiday film *It's a Wonderful Life* is, among other things, the story of a clash between Darwinian angels and Darwinian demons. The angels are represented by the film's main character, George Bailey, who has spent his life in service to his friends and neighbors in his hometown of Bedford Falls. In game-theoretical terms, Bailey consistently displays behavior that may at first glance look irrational, given that he is forever forsaking his own advantage for the greater good of the group. In colloquial terms, Bailey is the ultimate mensch.

If George Bailey represents the angels, then the Darwinian demons take the form of the film's antagonist, Mr. Potter, the richest and meanest man in town. Potter has spent his life maximizing his own advantage at every possible opportunity, even though doing so has degraded Bedford Falls for everyone else. Potter sees Bailey as an obstacle to his victory in the resources arms race—and so he connives to plunge Bailey into serious financial trouble in order to get him out of the way.

Faced with these money problems, Bailey contemplates suicide, only to be visited by an actual angel (of the heavenly variety), who gives him a chance to see what the world would have been like had he never been born. To make a long story short, in Bailey's absence, Darwinian demons have overrun Bedford Falls. Without Bailey around to promote cooperation, the town has remade itself in Mr. Potter's image, and everyone there is focused on pursuing their own advantage. With a renewed sense of clarity about the value of his life's choices, Bailey abandons all thoughts of suicide and goes home.

Bailey is greeted there by dozens of his neighbors, who, upon hearing that he is in trouble, came together to raise enough money to solve his problem. After a lifetime of making unselfish choices that benefited others, Bailey's magnanimity is finally rewarded when the group decides en masse to make an unselfish choice of its own. At the end of the film, Bailey is hailed as "the richest man in town"—if not in terms of money, then surely in terms of good will.

So why did I just spend a page and a half recapping the plot of an eighty-year-old movie? Because George Bailey's cinematic redemption is a textbook example of the power of *indirect reciprocity*. The townsfolk came to Bailey's aid not because they expected to profit by doing so, or because they expected favors from Bailey in return, but because Bailey had a reputation of being a good guy who always helped others when they were in need. The people of Bedford Falls would not have been so generous for just anyone. If mean old Mr. Potter had found himself in deep debt, for instance, his neighbors would probably have thrown a party instead of coming to his assistance. But since Bailey was known for doing good works, it became rational for the townsfolk to indirectly reciprocate his kindness by choosing to cooperate and help repay his debt.

The concept of *reputational scoring* is central to the inner workings of indirect reciprocity. In one way or another, most of us informally keep track of which individuals in our environments model and abide by the community's values, and which ones choose to flout those values. (You can think of this process as a cousin to Santa Claus's "naughty or nice" list.) These implicit reputational scores not only influence how we feel about these various individuals; they influence the depth and quality of our interactions with them—and the extent to which we might be willing to disadvantage ourselves in order to help them.

There are evolutionary reasons for why our reputations tend to matter so much to us. Thousands of years ago, when most humans were nomadic hunter-gatherers, it was evolutionarily advantageous to belong to a tribe, because tribes offered resources and protection at levels unavailable to unaffiliated nomads. To be expelled from a tribe and be made to fend for yourself in this harsh, precivilizational world was often a literal death sentence. As such, it was extremely adaptive for individuals to cultivate respect within these groups, to abide by their standards and customs, and to avoid developing a reputation, for example, as someone who routinely steals other people's food.

While society has evolved since the days of ritual banishment, reputation remains a valuable ordering mechanism within modern communities. Imagine that one of your neighbors is always playing loud music late and leaving garbage everywhere, while another is quiet and tidy and loves to bake cakes for people. Which neighbor is more likely to win the support of the community if they fall on hard times and need a helping hand?

We use these methods of implicit reputational scoring to guide other everyday decisions, too. We might avoid businesses that are known for cheating or pressuring their customers, and patronize those known for their honesty. At the cinema, we might forgo a movie starring an actor who has been accused of sexual misconduct and gravitate instead to one whose star has a sterling public image. This implicit understanding that there can be tangible benefits to having a good reputation—and tangible consequences for having a bad one—can serve to encourage cooperative behavior while discouraging defection at times when defecting might otherwise be the rational move.

In the last chapter, we discussed centralized solutions to the existentially risky problems that currently plague the world—and determined that we're still quite far away from plausibly achieving a viable world government. On the other hand, the principle of indirect reciprocity, as embodied in these implicit social mechanisms for reputational scoring, offers a *decentralized* answer to the question of how we might lower the world's existential risk levels—and perhaps even vanquish Darwin's demons once and for all.

It is probably clear to you how reputation is decentralized. While a centralized global government can enforce cooperation on a top-down basis, no one single entity can determine a person's reputation. George Bailey and Mr. Potter didn't get their good guy–bad guy reputations because the mayor of Bedford Falls proclaimed them as such. Rather, their reputations arose from repeated interactions with various people over the course of many years. Not only can decentralized mechanisms for cooperation be a bulwark against totalitarianism, they are also much easier to implement than centralized ones.

In the rest of this chapter, I will explain how we might take this simple concept of decentralized reputational scoring and expand it into a tool that makes it adaptive for countries, corporations, and other powerful global entities to cooperate rather than defect. Such a tool might become a very useful mechanism for encouraging entities to reject their inner Mr. Potter and instead follow the example set by George Bailey.

How Credit Risk Ratings Promote Cooperation

You may be familiar with the concept of a credit risk rating, which essentially quantifies an entity's reliability and trustworthiness in order to determine its eligibility for loans, lines of credit, credit cards, and other financial instruments. The credit risk rating—and its individual equivalent, the credit score—measures not just an entity's ability to pay on time and in full but its willingness to do so, as shown by its past history.

Credit risk ratings serve as an *explicit* reputational score, used to reward agents who choose to act cooperatively. This sort of explicit reputational scoring is used in other business contexts, too. On websites such as Trustpilot, Yelp, and TripAdvisor, customers can post reviews of businesses they have patronized. Taken cumulatively and adjusting for outliers, these websites serve as reputational custodians for public-facing businesses. The reviews published there spotlight those businesses that meet or exceed their obligations to their customers—as well as those that fail to do so. Businesses can publicly respond to these reviews, too, and the substance of those responses further clarifies these businesses' reputations.

Likewise, on websites such as Glassdoor, current and former

employees of companies can anonymously review their workplaces, thus creating an explicit record of businesses that treat their workers well, and of those that incubate toxic work environments. This information can be immensely valuable to prospective employees in deciding whether to start a career at a given company. All else being equal, employers might prefer to skimp on perks and salaries in order to increase their profit margins. But by making this review data public, Glassdoor creates an incentive for businesses to treat their workers well, lest they lose their ability to attract top talent.

These explicit reputational mechanisms, such as credit risk ratings and aggregated online reviews, also work on a decentralized basis. A credit risk rating isn't the result of one single imperious banker capriciously determining an entity's creditworthiness; rather, it's a distilled representation of that entity's financial interactions within many different contexts. A TripAdvisor ranking isn't the product of a single powerful travel writer proclaiming some places good and others bad; rather, it's aggregated from many reviews left by individual people over the years.

These reputational mechanisms have fundamentally changed the fitness landscape within their business contexts. Before credit risk ratings were in common use, for instance, lenders and investors had no easy, universal way to share reliable information that might illuminate an institution's history of default. This informational void worked to the benefit of unscrupulous institutional borrowers, which could work their way through one lender or investor after another—defaulting on each along the way—without worrying too much that their history of bad behavior would ever catch up with them. And in the era before online reviews, restaurants in tourist towns could more easily take advantage of visitors, serving overpriced food without having to worry about any blowback or long-term harm to their business.

In a world where the short-term survival impulse often motivates people and entities to rapaciously pursue their own advantage—even if doing so degrades their surrounding environments—public-facing, institutional reputational repositories can make it adaptive to cultivate a good reputation while also raising the cost of defection to the point where it is no longer a profitable choice. These explicit reputational mechanisms are not perfect ones, of course: they can be manipulated; they can be subject to Goodhart's Law; and at times they may seem to create as many problems as they solve. But I'd be willing to bet that, given the choice, most people would not want to return to a world without them.

As an example of how even the strongest, most storied entities can crumble if their reputations sink beyond salvage, we need only consider the fall from grace of the blue-chip auditing firm Arthur Andersen. This firm, celebrated for its integrity and a cornerstone in the global financial auditing sector, stood tall as one of the "Big Five" accounting giants. Its prestige and reliability were unquestioned, built over decades since its founding in 1913. But the winds of fate shifted dramatically with the infamous Enron scandal in 2001.

As the trusted auditor of Enron, a Texas energy company that used unethical accounting methods to conceal debt while feigning runaway profitability, Arthur Andersen found itself entangled in one of the most scandalous corporate frauds in history. Its failure to detect and report Enron's deep-rooted financial manipulations was just the beginning of its woes. The situation escalated when investigators learned that Arthur Andersen employees were shredding tons of documents related to their Enron audits—an incriminating attempt to bury the truth.

The news sent shock waves through the business world. Clients hurriedly distanced themselves from an auditor now synonymous with scandal and deceit. The once-revered name of Arthur Andersen became a byword for corporate malfeasance. The collapse was as swift as it was stunning. In a blink, a firm that had weathered storms for nearly a century had crumbled.

The Arthur Andersen example shows how reputation can be a powerful determinant of an entity's fitness in its environment. What if we applied the same concept to addressing the major existential risks that the planet faces today? Is it plausible to think that the straightforward principles of reputation might be used to prevent selfish global actors from doing things that bring the world closer to ruin? To answer that question, we must first examine the interconnected nature of the world economy.

The Power of Supply Chains and the $1,500 Chicken Sandwich

Our world economy is essentially a large network where materials, labor, and capital flow between entities. (These flows are also called *value chains* or *supply chains*. I will often use these terms synonymously.) In order to survive and thrive in the world economy, an entity must have access to these flows. Even in the simplest cases—your local sandwich shop, for instance—not having access to these flows is basically a death sentence, much as, for early humans, being shunned by the tribe used to be a death sentence.

To illustrate this point, consider the simple act of making a chicken sandwich, which you can buy for a few dollars at your grocery store. You might think that making such a sandwich is a trivial task: you buy the chicken, the bread, the sauce, and the pickles, you put them all together, and you're finished. But that's not really making a chicken sandwich from scratch. To truly make it from scratch, and not rely on someone else's work, you would have to grow the lettuce, pickle the cucumbers, feed the chicken for months and slaughter it, grow and mill the grain for the bread, discover the secret behind the secret sauce, and so on. Andy George, a YouTuber running the channel *How to Make Everything,* tried to do just that in a video titled "How to Make a $1500 Sandwich in Only 6 Months." In other words, without the benefit of specialized global supply chains, making a simple sandwich is one thousand times more expensive, and takes months to complete.

Or take a smartphone, which exists on an entirely different plane of complexity than a chicken sandwich. It is safe to say that there is not a single individual in the world today who could build a smartphone truly from scratch—mining precious minerals, creating the lithography machines, designing the circuits, and building the transistors and motherboards, assembly, and operating system. It is exactly this technological dependency that makes our societies so very vulnerable. We are all reliant now on these global supply chains to underwrite the conveniences of life in the modern world. If these value chains were to fall apart, then soon so would we.

The vulnerability of these global value chains was illustrated in early 2021, when the *Ever Given,* a massive container ship, got lodged in the Suez Canal, one of the world's most crucial waterways. The ship was stuck there for nearly a week, during which time hundreds of other vessels were left waiting to pass through the canal. This logjam caused a massive disruption in global supply chains, leaving an estimated $10 billion worth of trade stranded.

This vulnerability can also be beneficial in certain contexts. If producing a simple chicken sandwich is a monumental undertaking, then surely producing dangerous technologies and weapons is exponentially more difficult. As of now, with AI still in its nascence, none of these existentially risky technologies can be developed in a vacuum by single actors. Instead, they all depend on coordination and collaboration within global value chains. Every single new product, technology, and breakthrough represents the sum total of a massive, distributed collaboration between different actors in the value chain—and therein lies our opportunity to reset the incentives and slow the pace of our own destruction.

Every supplier of critical materials or necessary subcomponents, every irreplaceable highly skilled worker, every large-scale investor, every critical transportation route—every one of them constitutes a potential *governance node* in the value chain of any given actor. If any such node chooses to stand up and say, "I will not cooperate with you unless you behave more responsibly," then they can slow down, and in some cases entirely stop, an actor's risky or unethical action. The image on page 198 depicts a typical value chain, in which each black dot symbolizes an entity, and each connecting line represents a trade relationship. Certain actors act as bottlenecks in the network, represented by white nodes, highlighting their significance as potential governance nodes.

For a real-life example of this process in action, we need look no further than the execution chambers of the US prison system. For decades, in states where capital punishment was legal, lethal injection was the preferred execution method. About fifteen years ago, though—in part due to pressure from opponents of capital punishment—the pharmaceutical firms that manufactured these lethal-injection drugs began either to take them out of production

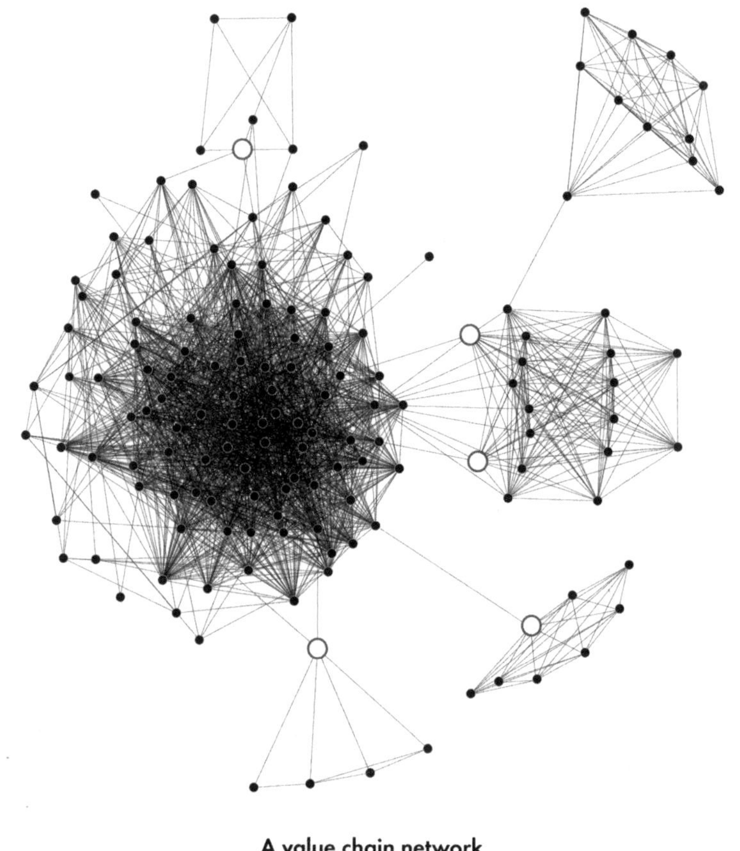

A value chain network
(governance nodes are white)

or to restrict their use for execution purposes. As a result, executions slowed to a halt in many states, an outcome exemplifying the power held by the nodes in these value chains.

The governance nodes in existentially risky value chains can wield similar power. For instance, building nuclear weapons requires uranium and the procurement of highly specialized machinery. Likewise, building AI requires highly advanced GPUs that, as of this writing, are so scarce—the company Nvidia and their manufacturer TSMC holds a near-monopoly on their production—that they've been likened to rare-earth metals. Exploiting resources such as gas and oil likewise requires sophisticated equipment and billions of dollars in investments. In other words, we do not yet live in a world where a single rogue player can create a doomsday weapon without at least the indirect complicity of many other employees, investors, and suppliers from a multitude of countries, either by buying the right machines or equipment from them, or by getting investment from them.

Interconnected value chains equip numerous actors with a formidable tool. This tool is not potent enough to shatter the world, but it is sufficiently powerful to halt those who might unwittingly threaten its very existence, or the values that we hold dear—especially if the relevant actors can learn to coordinate with one another. In order to govern a dangerous technology or entity, we just need to find the right governance node in the value chain. We can turn the problem of AI governance into the value chain governance of computer chips. We can turn bioweapons governance into the governance of critical lab equipment. We can turn nuclear weapons governance into the value chain governance of uranium. We can turn climate change governance into the governance of fossil fuels. This insight has long been utilized by governmental bodies. It spans from the International Atomic Energy Agency's oversight of uranium enrichment activities to the United States' recent imposition of export controls on advanced computing hardware to China, and the restrictions on the distribution of hazardous chemicals and laboratory equipment.

The recent invasion of Ukraine has underscored the significance of sanctions rooted in reputational concerns, while also revealing some limitations. Nation-states weren't the only entities that responded to Russia's incursion: more than one thousand companies have publicly declared their decision to scale back operations in Russia, cutting ties with Russian firms and subsidiaries to mitigate reputational risks. Although most of these withdrawals had minimal impact on Russia's economy or its military efforts, certain US and European suppliers were pinpointed as critical governance nodes that could significantly disrupt Russian armament manufacturers. Modern military hardware requires the extensive use of computer chips, the bulk of which are produced by US companies. Consequently, stringent sanctions were imposed on US chipmakers to halt the supply of these essential components to Russia.

Still, nefarious actors often find ways to navigate around sanctions. Despite heavy sanctions, 50 percent of the electronic components in Russian guided missiles are produced by the United States; when it comes to Iranian missiles, this figure rises to 80 percent. These actors bypass sanctions through intricate networks of intermediaries, such as companies in Singapore, the Maldives, Germany, and elsewhere, which purchase the components under false pretenses and then resell them to Iran and Russia. How can any entity, especially a critical governance node, effectively sanction actors with a low reputation if those sanctions can be so easily circumvented?

There are several concrete measures any entity can use to thwart these circumvention efforts. Immutable recordkeeping facilitated by blockchain technology provides a secure, unalterable history of transactions, enabling precise tracking of ownership of these risk-enabling components. For electronic risk-enabling components, the integration of connected sensors allows for real-time monitoring of component locations and usage, further enhanced by biometric locks to restrict access to authorized individuals only. Nonelectronic risk-enabling components can be monitored through randomly selected on-site audits, ensuring these components reach and remain at their intended destinations. These technologies are not new, and they can be deployed cost-effectively. Consider the traceability of motor vehicles in many developed countries, where public ownership registries are regularly updated, independent audits of vehicle functionality are conducted annually with ownership verification, and the location of vehicles can be easily tracked via GPS.

I hope I've shown that, at least theoretically, a decentralized reputational scoring system leveraging the distributed nature of global supply chains could be a powerful mechanism to incentivize global cooperation. But the sorts of actions that earn you a favorable reputation in the business world do not always align with core human values, which prompts another question: Is it possible to create a world where the reputation that people care about more directly matches the amount of good the entity does for the world?

In this world, an entity's reputation would be tarnished if it engaged in behavior that might lead to global catastrophic risks. Few would want to buy from a company with a bad reputation; it would be nearly impossible for that company to get insurance, investments, loans, or talented employees. In such a world, it would be adaptive for any actor to lean toward the white balls, instead of toward the black ones that optimize their hunger for resources and power. How might we devise an empirical system to meaningfully measure an entity's reputation in such a way that might promote positive global change?

For those with some knowledge of the domain of finance and corporate responsibility, the concept of Environmental, Social, and

Corporate Governance (ESG) ratings will immediately come to mind. Essentially, their purpose is to provide buyers and investors with reputational data to mitigate risks in their commercial value chain relationships. Let's delve into whether these ratings are effective in establishing a reputational system capable of fostering decentralized cooperation and policing defection.

A Case Study in Misaligned Metrics

ESG ratings started from a good place. They were built around a document called the United Nations Principles for Responsible Investments (PRI), which was meant to encourage companies to engage in "conscious capitalism" and focus on more than just quarterly profits. A staggering 90 percent of the assets managed by global funds—amounting to $121.3 trillion—adhere to the PRI today.

As ESG grew in popularity, financial data brokers saw an opportunity to sell investors on a new type of product called ESG ratings, loosely grounded in the aforementioned principles. But how do you go about compiling qualitative principles such as climate impact, diversity, and human rights into a quantitative score?

The data brokers didn't quite know how to do that, so every ESG rating agency added its own more or less subjective weight to each qualitative component. For example, Refinitiv might assign a weight of 0.002 to greenhouse gas emissions, while Sustainalytics might assign it a much higher weight of 0.048—twenty-four times greater. Similarly, MSCI might allocate a weight of 0.388 to corruption, in contrast to Moody's 0.072, which is 5.4 times lower. Researchers found that the average correlation in ratings for the same companies across different ESG providers was only

54 percent, with some as low as 13.9 percent—indicating virtually no correlation at all. In comparison, credit ratings from S&P, Moody's, and Fitch exhibit correlations between 94 percent and 96 percent. This stark difference in consistency is highlighted in the following illustration.

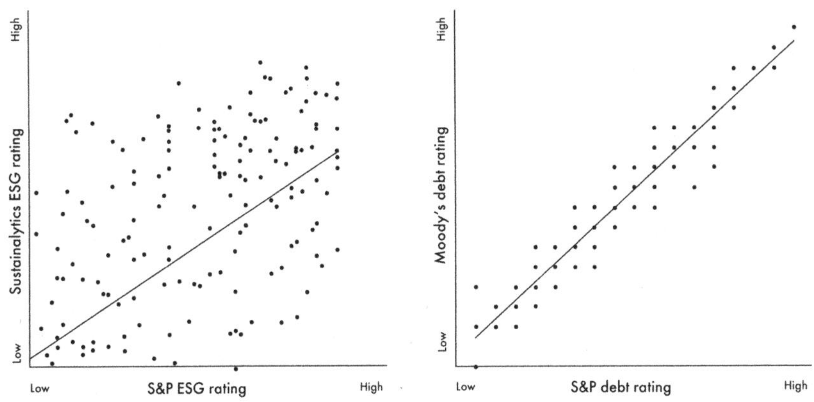

Perhaps more concerning is that the weights in ESG ratings tend to prioritize policies, targets, processes, and promises over tangible outcomes. Are you seeking a higher ESG score without having to enact substantial change? In many cases, the solution is, unfortunately, straightforward: produce an abundance of impressive-looking documents discussing at length the potential impacts of future climate change on your business, while still doing little to mitigate climate change in the here and now. The relative ease with which ESG ratings can be gamed by companies can create somewhat paradoxical results. For example, the electric car company Tesla has a lower ESG score than the oil company Shell.

The rise of ESG metrics once again brings us back to Goodhart's Law, which suggests that when a measure becomes a target, it ceases to be a good measure. (In this context, though, it seems like the measures in question were never good ones to begin with.) As of now, attaining a high ESG rating has more to do with gaming a narrowly defined metric of success than with achieving meaningful outcomes. Just as it's often simpler to reduce reported crime statistics than to actually prevent crime, or to distribute textbooks in low-income countries than to enhance learning outcomes, it's less challenging to generate paperwork that results in a higher ESG score than to genuinely reduce emissions or develop products that positively impact the world.

Despite these shortcomings, I still believe that the broader ESG movement has had a net positive impact on the world. There is nothing inherently wrong with encouraging countries, corporations, and other powerful entities to act in ways that align with basic human values; nor is there anything wrong with finding ways to make it unprofitable for these entities to commit socially and environmentally harmful acts. However, for these principles to translate into true, globally transformative impact, we would need to update our reputational ratings in a way that defeats the Darwinian demons embedded in our current system by overcoming Goodhart's Law.

In the next chapter, I'll go deep into the principles around which such a score might be built. For now, I'll note that rather than assigning arbitrary weights across various domains, we must structure our reputational score in such a way that it is more costly to manipulate the score than to genuinely improve it through actions that lead to positive, intrinsically valuable outcomes. Furthermore, it's crucial to design our scoring system in a manner that obviates the need to rely on the subjectivity of a centralized authority for assigning these weights. Instead, we should adhere to a well-established principle in computer security: *never trust, always verify*.

This principle underscores the importance of decentralizing trust, as doing so requires the weightings of the score to be grounded in scientifically measurable indicators. Decentralization ensures that, in the absence of a central authority, participants would obtain the same measurable outcomes if they themselves replicated the same procedure. For instance, the circumference of the Earth was independently measured and verified by both Greek and Islamic scholars; in the same way, we would want our reputational score to be independently replicable.

If we had a well-functioning reputational score, then markets would transform into the most protective forces in society, safeguarding not only our present interests but also our future well-being and the sustainability of our planet. In this world, we would lift the market by its bootstraps and make it a tool for aligning economic activities with what we intrinsically value as a society. Such a score would foster a culture where the reputation of an entity is measured not only by its financial performance but also by its contribution to averting existential risks and promoting a sustainable, thriving future. The market, traditionally seen as a driver of competition and growth, would evolve into a mechanism for fostering cooperation, foresight, and responsible stewardship of the planet.

How do we actually go about creating such a score? It starts with taking a new look at one of the oldest practices in human civilization: accounting.

CHAPTER ELEVEN

Measuring What Matters

Since the earliest days of settled civilization, humans have sought new and better ways to measure and keep track of the things they value. The entire notion of wealth, after all, relies on the ability to tabulate one's worth and measure it against another's. Accounting measures growth and decline, the forward and/or backward progress of companies and kingdoms and the individual agents therein.

As early tribes grew into chiefdoms, states, and even empires, their accounting systems adapted accordingly. The Roman Empire's expansive needs for infrastructure such as roads, sanitation, and military defense necessitated a sophisticated accounting system. This system included meticulous records of assets such as land, slaves, livestock, and other forms of wealth, with taxes levied accordingly. Similarly, advanced accounting systems emerged in ancient civilizations as disparate as Mesopotamia, Egypt, China, and the Indus Valley, reflecting each civilization's unique societal and economic structures.

In the fourteenth century, Florentine merchants improved on existing methods of accounting when they invented the record-keeping process known as double-entry bookkeeping. By using one part of a ledger entry to measure debits, and the other part to measure credits, double-entry bookkeeping offered a newly precise way of tracking profit and loss by "balancing the books." Instead of just noting down the profit from a sold carpet, the merchant would now note both the increase in his assets by the amount of the profit (in debit) and the source of that increase (revenue from a sold carpet, in credit). An unbalanced book—a ledger that documented more debits than credits, for instance—implied mismanagement, or perhaps fraud and malfeasance. Precise double entries made it easy to identify where the losses and revenue were coming from and who might be responsible for the potential fraud.

But traditional accounting, with its keen focus on monetary transactions, is just the tip of the iceberg when it comes to keeping track of what we value. Throughout this book, we've repeatedly observed that profit and loss are relatively narrow measures of success—and when we focus on narrow metrics to the exclusion of most other values and priorities, we can end up producing unintended negative consequences for the rest of the world.

Just as we assign monetary values to everyday goods and services, we should, at least in principle, be able to quantify a company or a government's impact on health, nature, and well-being, and to incorporate those figures into our financial calculations. Imagine if we could integrate significant societal costs such as climate change and pollution, which claim millions of lives annually, into our profit and loss sheets. By rigorously accounting for the externalities created by our current Darwinian-style economic competitions, we would take a giant step toward executing the Great Bootstrap.

The notion of making it adaptive for businesses to stringently account for their social impact is more than just a pipe dream. The rise of ESG metrics, flawed as those measures might be, shows that it is at least possible to incorporate other factors into our efforts to evaluate corporate performance. My own experience at the helm of Normative, meanwhile, has shown me that such an accounting system can work.

For example, the carbon accounting software we have built over the past decade measures the climate impact of our customers' businesses. To calculate a company's carbon footprint, we use various data sources: the emissions created when the company produces different materials; the amount of material the company is using in its production; the number of miles the company's cars have driven and whether the cars are electric; and so on. While the exact arithmetic depends on the context, I believe that similar methods can be built for other externalities as well.

In this chapter, I will introduce what I believe is the next necessary step for accounting: *value alignment*. This concept is designed to make it adaptive for the world's most impactful entities to account for human values as thoroughly and as effectively as they now account for profit and loss. Since the specific methodology will differ depending on what we are measuring, I will not purport to offer an exact formula for this new accounting system. Instead, I will look at the big picture and propose three design principles for value alignment—principles that are applicable in any context.

Think of them as a detailed roadmap of the steps necessary to transform a Darwinian demon into a Darwinian angel.

The three principles are:

- Principle 1: Establish which outcomes we value.
- Principle 2: Predict which actions might influence those outcomes.
- Principle 3: Incentivize actions that comply with those outcomes.

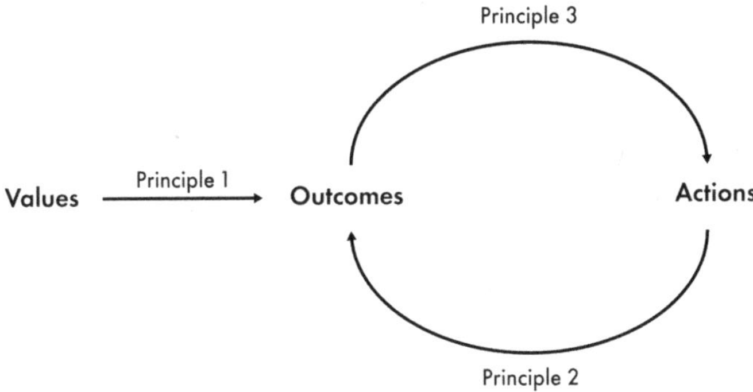

In a local setting, implementing this concept might be a relatively straightforward process. Indeed, many organizations already employ a similar approach. Take, for example, a newspaper in a competitive online environment that selects for insubstantial clickbait content. The paper's journalistic quality has gradually been declining, and the editors are seeking to reverse that trend.

To tackle the Darwinian demon plaguing the newspaper, the editors decide to examine the aforementioned principles. They begin with Principle 1, posing the questions "What do we intrinsically value?" and "How can we measure it?" After careful consideration, they unanimously recognize that their core value lies in holding power accountable. They decide to quantify this value by measuring the percentage of their articles dedicated to investigative journalism. This metric becomes a tangible reflection of their commitment to journalistic integrity and their foundational mission.

While Principle 1 focuses on defining the destination, or *Where do we want to go?*, Principle 2 is about charting the path: *How do we get there?* The editors gather for a brainstorming session, jotting down various strategic actions on a whiteboard. After hours of discussion, they realize that as long as the newspaper relies primarily on advertising revenue, it will always face pressure to churn out clickbait content that drives engagement with advertisers. Consequently, they identify the most crucial strategic action: diversifying their revenue model and launching an online subscription program, where premium content is available behind a paywall. If successfully executed, this shift in strategy will help the newspaper refocus on producing quality journalism, while freeing it from some of the constraints imposed by an overreliance on advertiser funding. The editors conclude that this approach aligns well with their ultimate goal of holding power to account.

The next step is to get the journalists on board with this new strategy. The editors do this by turning to Principle 3, which focuses on aligning individual motivations with the organization's strategic goals. So the newspaper introduces a bonus program as a tangible incentive. Under this program, journalists are rewarded for writing the sorts of articles that are compelling enough to prompt readers to subscribe when they encounter the paywall. By directly linking journalistic excellence with financial reward, the newspaper aims to foster a culture where quality and depth are both valued and incentivized.

But while implementing the principles on an organizational level is a manageable task, sustaining them within a competitive environment can be a daunting challenge. Take, for example, Jeff Bezos's acquisition of *The Washington Post* in 2013. Under his leadership, the newspaper has shifted its online strategy to a subscription model that rewards investigative reporting. However, despite this principled approach, the *Post* faces competition from other outlets, including those that prioritize clickbait journalism to boost readership and revenue. Indeed, after much initial success with its renewed focus on subscription-worthy journalism, in late 2023 the *Post* posted a $100 million loss for the year and initiated substantial buyouts and layoffs. These cuts will only make it harder for the *Post* to keep producing in-depth accountability reporting—an outcome that illustrates the difficulty of adhering to values in a market where others may not play by the same rules.

One lone cooperator cannot easily compel similar behavior from other autonomous actors in their environment. Thus, if we are to successfully achieve the Great Bootstrap and conquer Darwinian demons wherever they are found, we must find a way to promote the principles of value alignment on a global scale. To get from here to there, I propose that we establish a *reputational market*, largely inspired by Robin Hanson's idea of *futarchy*.

A reputational market would operate on similar principles to the stock market. Like the stock market, it would revolve around speculative investment—but it would emphasize companies' societal impact rather than their financial performance. In the stock market, investors buy and sell shares based on their predictions of a company's future financial success. In a reputational market, the trade is centered on predictions about a company's future impact on the world, be it positive or negative.

This market doesn't just create a pricing signal for the monetary value of a company, as seen in stock markets. Instead, it generates a *reputational score*. This score represents the market's collective belief in the potential good that a company will contribute to the world, based on the intrinsic values that we hold dear. In the following three sections, I will outline how we might establish such a marketplace and ground it in the three universal principles of value alignment.

Principle 1: Establish Which Outcomes We Value

In late 2023, while speaking on a panel at the United Nations COP28 climate conference, I told the audience that I didn't give a shit about carbon emissions.

It was a true statement, albeit a somewhat misleading one. I was trying to communicate a broader point: if our concern for greenhouse gas emissions is based solely on their environmental impact, then why don't we express similar concerns about the emissions on Venus? The answer lies in the fact that on Venus, there are no sentient beings whose well-being is directly affected by these changes. My point was that the true intrinsic value lies not in the emissions themselves, but (at least for me) in their consequences on well-being. If a godlike AI were to be tasked with reducing carbon emissions worldwide, it might choose to achieve that goal by, say, killing all humans—a successful outcome by emissions-reduction standards, but a suboptimal one in terms of well-being.

This sort of misalignment between what we measure and what we value isn't limited to environmental issues. Indeed, the phenomenon pervades all levels of society. While gross domestic product is often used as a measure of a country's success, the metric fails to account for individual welfare, which might be the outcome we intrinsically value most. Philanthropic efforts to promote literacy often measure outputs such as the number of books distributed rather than intrinsically valuable learning outcomes. The first step toward value alignment is to get rid of these narrow success metrics. But what should we replace them with? The answer is simple: we should replace them with our intrinsic values.

An *intrinsic value* is something that is cherished for its own sake rather than as a means to another end. To identify whether an outcome is intrinsically or instrumentally valuable, it helps to ask a series of *why* questions. Consider someone volunteering at a local food bank. If asked why they volunteer, they might say that they want to help people in need. Further questions may uncover a deeply ingrained belief in human dignity and equality. Eventually, this line of questioning will lead to a value that cannot be dissected further: the intrinsic value.

The most extensive survey of intrinsic values I've found was conducted by Spencer Greenberg at ClearerThinking.org. This survey, known as the "intrinsic values test," involved more than 2,100 participants. It revealed that the most significant intrinsic value for participants was *That I have agency and can make choices for myself,* with 37 percent of respondents rating it as very important. The second most significant valued aspect was *That I continue to learn new things,* with 35 percent considering it very important. Interestingly, *That people all around the world don't suffer* ranked only fifteenth, with 19 percent of respondents ranking it as really important.

Questioning a regular volunteer at a local food bank.	**Questioning someone who buys a fashionable piece of clothing.**
First *why*: Why do you volunteer at the food bank? Response: "Because I want to help people in need." Second *why*: Why do you want to help people in need? Response: "Because I believe everyone deserves to have their basic needs met." Third *why*: Why do you believe everyone deserves to have their basic needs met? Response: "Because it's a matter of human dignity and equality." Fourth *why*: Why do you value human dignity and equality? Here the individual might realize they can't explain further. This indicates that human dignity and equality are intrinsically valued.	First *why*: Why are you buying this trendy piece of clothing? Response: "Because it's stylish and popular right now." Second *why*: Why do you want to wear something that's stylish and popular? Response: "Because it makes me feel confident and accepted by my peers." Third *why*: Why is feeling confident and accepted important to you? Response: "Because it helps me feel connected and part of a community." Fourth *why*: Why do you value feeling connected and being part of a community? Here the individual might realize that the core value driving their decision is a deep-seated need for belonging and social connection.

You might have an idea of what your own intrinsic values are, as well as what other people's intrinsic values seem to be, at least according to one study. But which of these intrinsic values should we choose to guide our societies? Is it even possible to identify a set of intrinsic values that are objectively most important? After all, a fitness nut might claim that health is the only fundamental value, while a hedonist might argue that pleasure is the most important thing. Even when two people agree that *both* health and pleasure are fundamental values, they might still differ on the relative importance they assign to them.

How can we bridge the divides created by fundamental value disagreements? A debate over the exact height of the Eiffel Tower can be straightforwardly resolved by consulting an encyclopedia.

More intriguingly, this issue can also be addressed independently of any central authority—simply travel to Paris and measure it directly using a ruler and basic trigonometry. But value statements stand in stark contrast to empirical facts. For intrinsic values, objective measurement tools comparable to a ruler are conspicuously absent, and any self-proclaimed authority who claims the arbiter's role in these value disputes should be met with a healthy dose of skepticism.

Yet in determining the height of the Eiffel Tower, a third option exists: the wisdom of the crowd. Soliciting estimates from a large group about the tower's height can reveal a telling pattern, with the accurate figure often receiving a statistically significant number of guesses—a concept familiar to viewers of *Who Wants to Be a Millionaire?* in its "Ask the Audience" lifeline. This principle can help us to reconcile our differences over fundamental values.

Much as we elect government officials by voting, surveys could be used to identify humanity's intrinsic values. To add specificity to each intrinsic value, participants would be prompted to also propose a metric that most accurately reflects the realization of that value. For example, one participant may identify well-being as their intrinsic value, and suggest that the most effective way to gauge the fulfillment of this value is by rating their personal well-being on a scale from one to ten. Another might prioritize freedom as their intrinsic value, proposing the Freedom House index as the most suitable metric to assess whether this value is being met. Should there be no existing measure that adequately captures their intrinsic value, participants could propose potential methods for its quantification. The outcome of such surveys could be thought of as humanity's *collective value function,* encapsulating our collective desires and preferences.

This collective value function is designed to be relatively resistant to Goodhart's Law and specification gaming, thanks to its inclusion of multiple measurable metrics for each intrinsic value, which serve as control metrics. For example, if reported health were the sole metric for the value of physical well-being, then it might be easy for an entity to persuade individuals to claim they are healthier than they are, rather than undertaking the measures necessary to genuinely improve health. However, by incorporating additional control metrics—such as accessibility to healthcare; psychological evaluations through regular, anonymous mental health surveys; and physical health indicators such as average life expectancy and rates of lifestyle-related diseases—the system becomes much more robust and resistant to manipulation. This multimetric approach ensures a more comprehensive and reliable assessment of each intrinsic value, making the system significantly more difficult to exploit.

Embarking on the ambitious project of identifying humanity's intrinsic values and correlating them with objective metrics might initially seem like an impossible task. How do we engage the entirety of humanity? Is it reasonable to expect individuals, many of whom may have never contemplated their core values deeply, to provide articulate responses? Furthermore, how can we expect non-experts to link their values with valid, quantifiable metrics?

However, history shows that daunting tasks have not always deterred human progress. For instance, the creation of the first precise map of France, utilizing what were at the time cutting-edge trigonometric tools, was initiated in the 1730s by the French astronomer Jacques Cassini and furthered by his lineage, notably by his son César-François Cassini de Thury (Cassini III) and his grandson Jean-Dominique Cassini (Cassini IV). This monumental task spanned fifty-six years. Likewise, the initial calculation of the national GDP in the United States during the 1930s represented a formidable challenge, requiring the manual aggregation of millions of business and tax records. History is replete with examples where the scale of the challenge did not impede the pursuit of significant achievements.

In modern times, the feasibility of such a project is significantly enhanced by technology. Currently, over 5.3 billion people have internet access; through strategic random sampling, a representative segment of the population could be efficiently identified. Additionally, the potential of interactive AI chat assistants could revolutionize this process. Such technology could help participants to articulate their intrinsic values and align them with appropriate metrics through guided conversations. This approach not only democratizes the process by making it accessible to a broader audience, but also leverages the power of digital tools to automate and streamline the data collection and analysis phases.

A crucial advantage of this approach is that it does not rely on a centralized authority to conduct these surveys with integrity. Theoretically, any party capable of investing a few hundred thousand dollars could independently conduct such a survey, offering a means to verify the results of any other survey. This process, in contrast to the arbitrary weighting and weak correlations often seen in ESG ratings, would foster robustness and convergence in findings. By enabling independent verification, this method ensures greater transparency and reliability in the assessment of collective values or opinions.

So how might this collective value function be applied practically in a reputational market? Let's imagine a fossil fuel company—call it "BigOil Inc."—whose reputation is being calculated with regard to the intrinsic values of *nature preservation* and *health,* measured in tons of CO_2-equivalent emissions and years of healthy life, respectively. But identifying intrinsic values is only one part of the broader equation. For the system to work, we must also establish a clear link between specific actions by specific actors and determine how these actions might impact those intrinsic values. This consideration leads us to the second principle of the system.

Principle 2: Predict Which Actions Might Bring About Outcomes We Value

The world is a complex place, in which some value-aligned actions may have indirect and unforeseen consequences on other things that we value. Consider the goal of promoting economic development in the global South. While this goal is a worthy one, interventions aimed at human betterment can inadvertently lead to increased carbon emissions due to an increase in economic activities—an outcome that is broadly bad for everyone. This dilemma is not to say that we shouldn't promote economic development in the global South. Instead, it underscores the fact that the world is a big ball of complexity, and creating change in one context might produce negative consequences in another one.

While it is difficult enough to accurately name and measure our intrinsic values, it is even more difficult to accurately model the complex second-order effects of our policies and actions. Are such predictions even feasible? The book *Superforecasting: The Art and Science of Prediction*, by Philip E. Tetlock and Dan Gardner, suggests that they are.

Superforecasting explores the world of professional "superforecasters," individuals who have demonstrated a remarkable ability to predict complex, uncertain events. Their success challenges the common notion that the future is inherently unpredictable, and that humans cannot forecast the future with any consistency or accuracy.

The effectiveness of forecasting methods can vary significantly based on the context. In some scenarios, as noted earlier when I mentioned the quiz show *Who Wants to Be a Millionaire?,* the most reliable lifeline often comes from tapping the audience's collective wisdom by asking them to vote on what they think the right answer is. In scientific fields such as astronomy or weather forecasting, mathematical equations and scientific models might provide the most accurate predictions. In the realm of technology forecasting—especially for technologies that might enable engineered pandemics—a synergy between human and machine intelligence can be invaluable. AI can process vast datasets, from scientific publications to trend analyses, to identify potential risks and consequences of new technologies. Meanwhile, human forecasters can apply their judgment to interpret these findings.

One of today's most potent methods of forecasting is a mechanism known as a *prediction market.* A prediction market operates as a trading platform where forecasters can profit by accurately anticipating future events—ranging from election outcomes and sports matches to the likelihood of geopolitical developments. The strength of these markets lies in their indifference to the methodologies employed for making predictions, so long as those predictions prove accurate. Whether participants rely on complex mathematical models, advanced supercomputing, or a team of elite forecasters, the market remains neutral; success is solely determined by the accuracy of the predictions. This methodological neutrality in turn makes it adaptive for participants to use the most effective tools and strategies available for the specific phenomena they aim to predict.

So how do prediction markets work on a practical basis? A prediction market is a platform through which participants can buy and sell contracts based on the outcomes of future events. The price of these contracts fluctuates based on the market participants' collective belief in the likelihood of these events. Prediction markets can be used to forecast a wide range of outcomes, from election results to business trends or scientific discoveries. When the event occurs, the market resolves: contracts for the winning outcome pay out, while others become worthless. Traders make profits if they have bought contracts for the winning outcome at a lower price than they sell them for, or if they hold them until resolution. Conversely, they incur losses if they hold contracts for losing outcomes, or if they sell winning contracts for less than the purchase price.

Our proposed reputational market is basically just a specific type of prediction market, where our collective intelligence and collective values create a reputational score for a specific entity, as market participants continuously bet on how that actor's current or future actions might influence our collective value function. Previous occurrences are factored into forecasts, too. The market assigns higher probability to predictions that have previously occurred. For example, if a certain US senator has won reelection in her state for five consecutive terms, then the market would likely deem it very probable that she will win reelection once again.

Here's how this process might work in the context of a reputational market. First, a series of questions for each entity are posted on the market by its participants. The specific questions deemed relevant by the reputational market will vary depending on the nature of the entity being evaluated. In the case of our theoretical oil company, a pertinent question might be "How many tons of CO_2-equivalent emissions will BigOil Inc. release in 2025?" For a mining company, a relevant inquiry could be "How many deaths will MegaMining Inc. cause due to unsafe working conditions in 2025?" For a company involved in distributing potentially hazardous technologies like CRISPR, an even more critical forecast might be "Will UltraCRISPR Inc. contribute to a global catastrophic risk that could threaten all future life in 2025?" Naturally, if an entity is expected to pose an existential risk, then its score would significantly decrease, since it will likely influence all other values in our collective value function.

Second, market participants engage in bets where they believe their domain expertise gives them an advantage, aiming to outperform the average predictions of others. For example, to predict the tons of CO_2-equivalent emissions that BigOil Inc. will release in 2025, participants select a specific number or range of numbers. They then decide how much money they wish to wager on each prediction.

But the market's belief is not represented by a single figure. Instead, it encompasses a range of probabilities for different outcomes. There might be a 5 percent likelihood of emissions surpassing 5 million metric tons, and a 90 percent chance that emissions will not fall below 0.01 million metric tons. By calculating the expected value—multiplying each possible outcome by its associated probability and then summing these products—the market derives a final prediction. For example, it might predict that BigOil Inc. will be responsible for emitting 0.3 million metric tons of CO_2-equivalent emissions, and could potentially cause 1053 years' reduction of healthy life due to pollution in all the urban areas where it operates in 2025.

Third, the market effectively consolidates the collective beliefs of its participants on various questions, thus offering a comprehensive perspective on anticipated outcomes. For instance, aggregated data might reveal a median forecast suggesting a 55 percent chance that BigOil Inc. will emit at least 0.25 million metric tons of CO_2-equivalent emissions.

Fourth, to calculate an overall score, we aggregate all the various components, each weighted according to its significance within the collective value function. For example, one thousand years of healthy living might be assigned an intrinsic value weight of 10 percent, and one million metric tons of carbon might receive a weight of 5 percent. Consequently, the calculation would proceed as follows: $0.3 \times 0.05 + 1.053 \times 0.1 = 0.1203$.

Thus far, the reputational score merely reflects the market's belief that BigOil Inc. is having a negative impact. However, it's important to recognize that this belief could ultimately be erroneous. In order for our reputational market to operate successfully, we would have to be able to verify the accuracy of predictions and reward accurate forecasters—hopefully begetting an arms race to make better and better predictions. Otherwise, we risk creating a strong incentive to manipulate the accounting system through wishful thinking or overly optimistic predictions. Participants with a large financial or political stake in the outcome of an event might attempt to influence the market by placing large, misleading bets. This practice, in turn, could distort the market's predictions, rendering them less reliable as indicators of future outcomes.

The Royal Society, one of the world's oldest and most prestigious scientific institutions, abides by the motto *Nullius in verba,* which is Latin for "Take nobody's word for it." In an ideal reputational market, the accuracy of a given prediction can be verifiable by anyone, eliminating the need to defer to any central authority. While there may not be a straightforward equivalent to "Just measure the Eiffel Tower" for moral questions such as "Should we prioritize nature protection?", there is a clear approach for empirical questions such as "How many tons of greenhouse gases did BigOil Inc. release?" This approach could involve independently auditing and fact-checking BigOil's climate reports.

But what if BigOil decides to never disclose its emissions in the first place? How could the predictions on the market then be resolved? A tactic long used by rating agencies to promote transparency involves awarding a better rating to an entity simply for showing transparency. This concept can be practically applied as follows. To motivate BigOil to share important information, market participants could vote on the disclosures they find most critical to the bet, with the weight of each vote possibly being proportional to the amount wagered. For example, the market might put a lot of weight on disclosures such as "publishing an independently audited greenhouse gas report for Q1" or "disclosing the capacity of the oil drilling equipment to be used." If BigOil chooses to release some of this information, it would receive a temporary boost in its reputational score, corresponding to the volume of votes the disclosure received. Conversely, should it decide against revealing information deemed relevant to the bet, it could face a temporary score reduction. This penalty could escalate as the bet approached its resolution date, further encouraging transparency.

For any of this to work, an entity must care about its score in the first place. Unless we change the incentive structure, however, most companies will continue to choose profit over reputation—which is why we must make it rational to play along.

Principle 3: Incentivize Actions Predicted to Bring About Outcomes We Value

For our reputational system to proliferate across global value chains, the cost of maintaining a high score must be lower than the benefit of possessing a high score. How might we make this happen?

The cost of "acting in a way that optimizes your reputational score" must be reasonably low, which means that the effort, resources, and changes required for an entity to improve its score should not be prohibitively high. Entities must be able to achieve and maintain a high score without incurring overwhelming costs. For some social dilemmas, such as having equal representation for men and women on a board of directors, it would be reasonably inexpensive to update your score. But when it comes to global catastrophic risks such as climate change, taking meaningful action can be quite costly indeed.

This unfortunate fact leads us to the second condition: the benefits of optimizing one's reputational score must be sufficiently high to make it worth doing. Entities must perceive a clear, tangible advantage in having a high reputational score. Here is where the power of the governance nodes (e.g., hard-to-replace suppliers or investors), discussed in Chapter 10, can play an instrumental role.

To see how this process might work, let's turn back to our example of BigOil Inc. BigOil's largest customer might be a local utility provider, Electricity Inc., which buys its oil to generate electricity for millions of homes and numerous industrial companies. If Electricity Inc. were to cut ties with BigOil Inc. or, even more drastically, switch to solar energy, it would have a significant negative impact on BigOil. This makes Electricity Inc. into a critical governance node in the BigOil value chain.

However, if the relationship is mutually beneficial, then why would Electricity Inc. ever consider such a move? Here is where an important incentive mechanism comes into play: if the reputational market is effective, then doing business with BigOil would decrease Electricity Inc.'s reputation, since its patronage would indirectly support BigOil's oil-drilling activities. Indeed, several systems—ranging from the web of trust in cryptography and Google's PageRank algorithm to mutual credit systems—evaluate an entity's standing based on the ratings of the entities to which they are interconnected.

But why would Electricity Inc. be concerned about its reputational score? It, too, might depend on a critical governance node, whose cooperation is indispensable. This situation may appear somewhat recursive, reminiscent of nesting dolls placed one inside another. For this system to function effectively, there needs to be a foundational governance node—akin to the innermost doll in a set of nesting dolls—that possesses sufficient influence to gradually shift each node from indifference to a genuine concern for its reputational score.

In the OECD countries alone—representing thirty-eight developed democracies worldwide—pension funds manage assets totaling $51 trillion, likely representing about half of the world's total assets. This staggering sum grants these funds considerable leverage in any value chain. Furthermore, pension funds often have fiduciary responsibilities to act in the best interest of the general public. This responsibility may make them more inclined to engage in practices that safeguard the interests of the public, including in areas such as climate change. In fact, pension funds are the core driver of the fact that 90 percent of the world's assets under management have signed up to the UN Principles of Responsible Investment, as I mentioned in the previous chapter.

For the reputational market to effectively punish governance nodes such as Electricity Inc. for doing business with low reputation suppliers, it is crucial that the market is aware of BigOil being a supplier to Electricity Inc. in the first place. This awareness requires corporate transparency, perhaps motivated by the incentives for disclosure highlighted in the previous section. For instance, the market could facilitate a wager concerning the years of healthy life that will be lost within the supply chain of Electricity Inc. A compelling way for companies to enhance their reputation under this model would be to publish a supply chain transparency report, thereby clarifying their business relationships and impacts.

The idea that certain entities within a value chain wield enough influence to affect the behavior of others is not merely a theoretical concept. Indeed, this principle is already applied in the realm of corporate emissions management, where achieving net-zero emissions requires that a company's suppliers also aim for net zero. Specifically, the field of corporate carbon accounting introduces the idea of Scope 3 emissions, which is broadly defined as the emissions originating from one's value chain. This concept means that for a company to reach its net-zero emissions target, all its suppliers to reduce their emissions must also aim for, and achieve, net-zero emissions. The Scope 3 emissions concept is already putting a lot of pressure on suppliers to reduce their emissions.

At Normative, we capitalized on the idea of governance nodes through the launch of what we call the Carbon Network. The premise was straightforward: to manage your emissions effectively, you must encourage your suppliers to join the software platform. Since its launch at the end of 2023, each of our clients, on average, has invited nineteen of their suppliers, culminating in thousands of suppliers on the platform actively working to reduce their carbon emissions. Atop that food chain, largely responsible for creating that demand, are large banks and pension funds.

These ideas can also apply to the AI industry's supply chain. Advanced AI development requires sophisticated computer chips, all of which are produced by TSMC in Taiwan. Yet TSMC depends on special lithography machines from ASML in the Netherlands to make these chips.

Consider a scenario where a hypothetical company, Irresponsible AI Inc., is believed to cause harm. Given Irresponsible AI's notoriety, TSMC might worry about its reputation and the risk it runs of losing access to ASML's lithography machines if it does business with Irresponsible AI. To prevent reputational damage, TSMC could stop supplying chips to Irresponsible AI.

But if TSMC ignores the warning and its reputation subsequently suffers, then ASML would face a dilemma. Continuing to do business with TSMC could harm ASML's reputation, thus jeopardizing its funding from major pension funds. This funding is crucial for ASML's development of future lithography machines. Consequently, ASML might pressure TSMC to cut ties with Irresponsible AI, in order to safeguard ASML's ability to innovate. Upon reaching an agreement, both companies' reputations are salvaged. At this juncture, Irresponsible AI must either alter its approach or risk losing out to more ethical competitors who would retain access to high-end computer chips.

Forecasting future global catastrophic threats wouldn't just incentivize the market to behave more responsibly—it could also incentivize the development of corresponding "antidote technologies." If developed first, these technologies can preemptively mitigate the original risks. For example, in the case of engineered pandemics, potential defenses include Far-UVC irradiation to eliminate airborne pathogens, advanced wastewater surveillance, and improvements in filtering and ventilation. In the realm of AI, one antidote technology would be to design hardware locks into AI chips that prevent the training of noncertified models over a specific size. This concept is known as "differential technology development," a method by which we deliberately prioritize the development of certain antidote technologies, based on forecasting and predictions, before advancing other potentially dangerous technologies.

In summary, these reputational markets could do substantial good for the world. Imagine if, before launching operations in Oklahoma, the Picher Lead and Zinc Company had been forced to find a more permanent storage solution for its toxic waste, because its lack of permanent waste disposal plans gave it a reputational score so low that investors didn't want to touch it. Imagine a world where reputational markets could assess the extent to which each large enterprise operates within our planetary boundaries, and could force outliers to adjust. Imagine if every government that started a dangerous weapons program knew that it would be sanctioned by all other countries, making it next to impossible for them to procure any of the supplies or technologies required for the program to be successful.

But not all demons are so easy to deter. Some phenomena are inherently unpredictable, and try as we might to safeguard intrinsic values and mitigate existential risks, reputational scoring alone might not be enough to save us.

Can Reputation Alone Create the Great Bootstrap?

The Turkey Problem, as elucidated by Nassim Nicholas Taleb in his seminal work *The Black Swan,* revolves around the experiences of a turkey being fed daily. For one thousand days with each feeding, the turkey's well-being improves, and it grows more convinced that the benevolent humans who feed it have its best interests at heart. And then, without warning, the turkey's well-being level suddenly drops to 0. (One assumes that day 1,001 was in late November.)

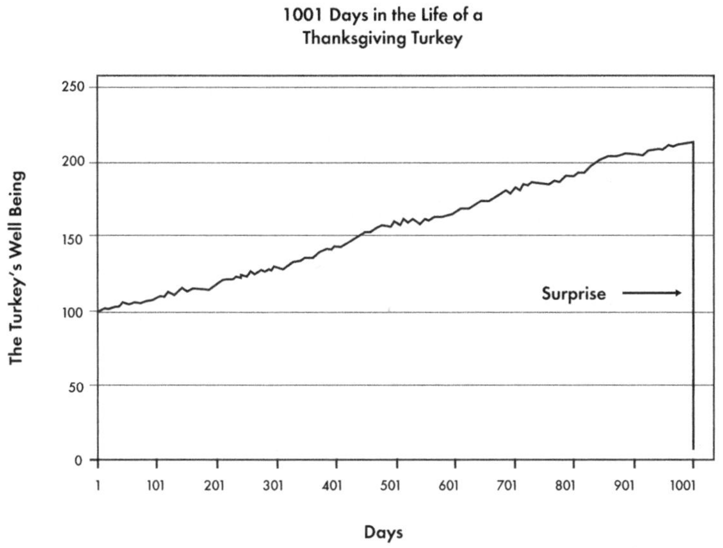

What if there are existential risks that we are intrinsically incapable of predicting, no matter how hard we might try? Contradicting

the idea of Tetlock and Gardner's *Superforecasters*, Taleb argues that predicting the future may be a far more challenging task than commonly assumed. He emphasizes that unpredictable, high-impact events are more common and influential than our standard forecasting methods suggest—and that, by their very nature, these events cannot be meaningfully prepared for or defended against. Each technological stride, while perhaps initially enhancing our well-being, might also carry hidden risks. In this scenario, the abrupt and unforeseen culmination of these risks could leave us as surprised and unprepared as the proverbial Thanksgiving turkey on its final day.

So who is correct about our ability to forecast the future: Tetlock and Gardner, or Taleb? I like to believe that existential risks, unlike the roll of a die, are predictable. If not, then the most effective way to avoid new technologies that might bring about unpredictably devastating consequences would be to halt all scientific progress—a strategy that is neither feasible nor desirable. Thus, we are left with the most practical course of action: attempting to predict the consequences of our technological advancements to the best of our ability. It's a delicate balance, navigating the need to promote progress with the imperative to anticipate and mitigate unforeseen risks. But this balance is also a vital one.

Building a reputational market will not be an easy task, and there will be plenty of other problems to solve along the way. What are the most robust control metrics to choose for any intrinsic value? How do we set up the market in such a way that complex predictions can effortlessly be resolved and yet independently verified by anyone with enough resources (at least in principle)? How should we estimate the value of future generations? How do we prevent chaotic interconnected market dynamics where everyone's score rapidly falls to zero? How do we aggregate occasionally contradicting values in a nonbiased way?

Moreover, it is hard to see how this decentralized approach might work in situations where a vertically integrated actor controls its entire value chain. Just as it is difficult to effectively sanction a fully self-sufficient actor, it would also be hard to use the threat of a diminished reputation to compel such an actor to cooperate. North Korea's journey to becoming a nuclear state underscores this phenomenon. Despite the significant obstacles presented by international sanctions, which prolonged the country's mastery of nuclear technology over five decades, North Korea's substantial domestic uranium reserves played a critical role in allowing the nation to develop those technologies anyway. These reserves enabled the nation to develop a nearly self-sufficient nuclear value chain, culminating in its emergence as a nuclear-armed state in 2006.

Is it possible to foster international policies to ensure that no single actor has enough power to own an entire existentially risky value chain? Do all critical value chains contain enough governance nodes that are sufficiently powerful to compel all participants to care about their own reputation as well as their neighbors' reputations? I suspect that the answer to these last two questions might very well be *no*. This is why I believe that, in order to implement a functional reputational market, we will ultimately need to deploy both centralized and decentralized mechanisms for promoting cooperation. The decentralized mechanisms based on reputation may be enough for the reputational market to gain traction. But we will also likely need centralized mechanisms based on force, as we described in Chapter 9, to equip that market to deal with *all* Darwinian demons. In fact, both mechanisms can gradually strengthen each other.

As previously noted, the United States and its allies, through companies such as Nvidia and TSMC, dominate both the design and manufacturing of high-end computer chips. They have imposed export controls to prevent Chinese weapon manufacturers and other government entities from accessing these chips. Conversely, China and its allies exert control over most of the minerals and raw materials that are essential for chip production. They could reciprocate by imposing sanctions on US chipmakers, potentially hampering the US's ability to advance its AI capabilities, which China views as a security threat.

In the near term, this scenario may escalate into a more intense trade war. However, the global AI value chain's distributed nature, encompassing everything from raw materials to finished products, diminishes the likelihood of any single player achieving absolute dominance. Over time, this interdependence could compel both sides to engage in negotiations, recognizing their mutual need to develop advanced AI technologies. This realization could pave the way for the implementation of the MAGIC proposal, or something like it—granting members selective access to advanced AI technologies, and thus progressively fostering the creation of a centralized global federation of states. This prospect, while speculative, explores how centralized and decentralized systems might interact in an optimal setting.

Are reputation markets the best way to implement the three principles of value alignment? The most effective method to discover this is experimentation. Luckily, all we need to start a reputational market is software and a team with experience in coding and marketing. The hardest problem will be how to gradually make reputational markets a part of our global institutions—and from there to make the Great Bootstrap happen. For that, we cannot rely on reputational markets alone. We need your help. Your role in bringing the Great Bootstrap to life will be the topic of the next and final chapter of this book.

CHAPTER TWELVE

What You Can Do

While Darwinian demons are formidable foes, they are not invincible ones. Time and again throughout history, we have found ways to successfully reset the incentives in their environments in ways that have made it adaptive for agents to cooperate rather than compete. The existential threats that we face today are very real, but the good news is that we already know what we have to do in order to address them. By embracing mechanisms for cooperation and choosing to consort with Darwinian angels, we can thwart the demons that would destroy us—and perhaps even defeat them entirely.

Implausible? Perhaps. But after spending so much of this book sketching out worst-case scenarios, I'd like to begin the final chapter by envisioning a *best*-case scenario. Let's imagine a world in which—whether by means of a global reputational market, a unified central government, or some combination of these methods and others—humanity has successfully reset the incentives and made cooperation the norm rather than the exception. In this world, the existential risks that imperil us today have been significantly reduced, and the arms races that compel agents to seek their own advantage even at the price of global calamity have been slowed to the pace of a nice brisk jog. While agents still might hope to get rich, smart, and powerful, they are strongly disincentivized from doing so in ways that ignore or violate certain intrinsic human values.

As we journey toward the horizon of this unprecedented future, we'll find that this best-case world is remarkably different from the world we know now. It's a world in which markets have been transformed into the most protective force on the planet, safeguarding the interests of all life.

In this future, the fulfillment of citizens' core values is robustly and anonymously measured in near-real time through advanced digital means. A spirited competition of sorts has emerged among competing accounting standards, each vying to capture what we value more accurately and robustly. The ones that succeed in this endeavor are adopted by impact funds, driving a global economic system that prioritizes human and planetary well-being.

Reputational markets, bolstered by advanced AI working with human superforecasters, meticulously analyze the strategy documents, budgets, and policies of every significant organization. These entities are rated based on their potential impact on what we collectively value. It's a competitive landscape in which innovative prediction engines, AI-human synergies, and superforecasters vie to deliver the most accurate predictions. Success in this arena is not just celebrated; it is rewarded, ensuring that the best minds and machines are continually motivated to refine their predictive capabilities.

Our reputational score has become the guiding light for actions beneficial to the flourishing of humankind. The institutionalization of these scores has been gradual but profound. A sufficiently high reputational score can lead to corporate tax breaks, for instance, whereas those institutions with the lowest scores lose access to essential resources and services. This system has created a powerful incentive for individuals, corporations, and governments to act responsibly and ethically.

Perhaps the most striking aspect of this new world is how human cooperation has evolved. There's no central force or dictator micromanaging every action; instead, each entity—be it an individual, a company, or a government—acts in harmony with its neighbors, guided by the "core DNA" encoded in the reputational market. This decentralized approach removes the need for an Orwellian surveillance state, replacing it with a system that naturally encourages positive behavior.

This system has laid the foundation for a highly decentralized global democracy. Decisions are made at a local organizational level, yet the overarching influence of the reputational market ensures that it's adaptive for organizations to do the right thing. In this world, the pursuit of power and profit no longer comes at the expense of the planet or its inhabitants. Instead, these drives are channeled toward nurturing and protecting life in all its forms.

As we stand in this new dawn, we will look back with pride at the path we've traversed. Our once-fragmented efforts to protect our planet will have brought about the next and final major evolutionary transition, producing a unified, efficient system that balances human aspiration with ecological and social stewardship. In this world, markets don't just trade in goods and services; they trade in trust, responsibility, and the collective future of all life on Earth. This world is one in which no rats have been stripped of their tails by profit-seeking rat catchers; in which no child is traumatized by bootleg *PAW Patrol* videos; in which every action, every decision, is a step toward a thriving, harmonious planet.

It's time to take action to make this world a reality.

The preceding chapters have largely focused on macroscopic organizational solutions to the existential problems that plague us. While these big-picture strategies are necessary if we are to align our selection pressures with our values, I realize that all this talk about reputational markets and global governance bodies might leave readers feeling a bit stymied. I suggested earlier that lone individuals cannot yet really do all that much to destroy the world without support from larger entities. By that same logic, though, lone individuals probably can't do very much to *save* the world on their own, either.

And yet this logic doesn't quite capture the power that we all have to make a difference in our everyday lives—not to mention the influence that those cumulative individual differences might plausibly exert on a global scale. The fact is that individuals can make a difference in the fight against Darwinian demons. The adoption of a reputational market based on value alignment principles is possible only in a world where humans choose to harness their intelligence and volition in order to consciously reset the incentives in their environments. If we can make that choice on a macroscopic level, then we can make similar choices on a microscopic level, too.

So what can *you* do to help bring about the Great Bootstrap? In this chapter, I will outline a global action plan by which we all can contribute. It goes like this:

1. Vote with your vote: Elect politicians who advocate systemic approaches for solving global coordination problems.

2. Vote with your wallet: Create demand for reputational markets, in order to encourage start-ups or existing ratings agencies to invest in the idea.

3. Create awareness: We are all affected by Darwinian demons in our everyday lives. A necessary step for change is to raise awareness of the problem.

4. Go demon hunting: Use our three principles of value alignment to slay the demons in your everyday life.

The plan begins at the ballot box.

Vote with Your Vote

Electoral politics is largely a game of scapegoating. Politicians who are constantly running for office often find it expedient to blame all the world's problems on the doings of bad actors, be they greedy corporations, foreign leaders, low-level miscreants, or simply members of the opposing political party. While these scapegoating strategies are adaptive to the selection pressures currently inherent to running for office, individual voters can help reset those selection pressures by rewarding those rare politicians who model cooperative behavior.

As voters, one of the best things that we can do is to loudly and reliably support politicians who promise to take global risks and intrinsic values seriously—both with our votes and with campaign donations. During election season, then, try to support politicians who advocate systemic changes rather than merely blaming individual "bad apples." Look for leaders bold enough to discuss measures such as a carbon tax, or those who acknowledge the potential global risks of AI if not managed promptly. Elect officials who are ready to reshape incentive structures for the greater good of humanity; encourage people who model this mindset to run for office themselves. Disincentivize xenophobia by choosing representatives who believe in collaborating with countries like China on governing hazardous technologies, rather than competing with them in an existentially perilous arms race. Vote for those who prioritize preparedness for future pandemics and recognize the importance of living within our planetary boundaries.

By organizing ourselves and loudly proclaiming our support for such policies, we can help to make it in politicians' best interest to implement them. This strategy is not a foolproof one, nor is it in itself sufficient to wholly offset the Darwinian demons that have affected the world of politics for so long. But it is at least a start, and ultimately, we have to start somewhere. There are also plenty of other things that politicians can do today that will help us all in the battle to combat global risk factors.

- Foster incentives within the financial sector to embrace the type of practices outlined in the previous chapter.
- Establish a dedicated office or political function focused on forecasting existential risks enabled by future technologies. This entity would detail how governance within value chains, transparency, and the development of reputational markets can serve as effective measures to mitigate these risks.
- Implement mandatory disclosure requirements that ensure value chain transparency, such as by introducing

digital product passports that track and log an item's provenance, and establish measurement systems that reflect the intrinsic values of citizens and voters.

- Champion multidisciplinary research initiatives aimed at discovering innovative strategies for global cooperation, incorporating insights from economics, psychology, game theory, and biology, thereby circumventing the pitfalls of Darwinian demons.
- Allocate innovation funds and organize competitions to encourage both new entrepreneurs and established entities to develop and engage in reputational markets.
- Let existential risk and AI governance be a wake-up call for the strengthening and expansion of global institutions, including regional unions such as the EU, and fortify the role of the United Nations in fostering international collaboration.
- Promote the adoption of taxes on activities and resources that are misaligned with our intrinsic values, such as carbon emissions and the overexploitation of finite resources.
- Exercise caution against survivorship bias, especially in the context of international relations and nuclear diplomacy, recognizing the complexities and unintended consequences of aggressive posturing.
- Be brave and engage in "conspicuous acts of kindness" toward rivals in the tit-for-tat game of global politics, changing a downward spiral of defection into an upward trend of cooperation.

Vote with Your Wallet

Creating reputational markets that gradually gain popularity and spread throughout value chains will not be an easy task. Their success will largely depend on the incentive landscape in which they are deployed. It is therefore essential that we provide incentives for these markets to be created—quickly. We can do this by leveraging our power as buyers or investors in the value chain network. Let's start with consumer power.

Imagine every purchase you make as a vote, a small yet potent endorsement of a company's ethos and impact. Before you part with your hard-earned money, pause and ponder: *Does this business mirror my values? Are its operations leaving a footprint I'm comfortable with?* Opting for companies that resonate with your principles isn't just an act of personal alignment—it's a statement. And let's not forget that amid our diverse values, the universal baseline remains: the survival of our planet. Our collective existence and the avoidance of global catastrophic risks hinge on our ability to operate within Earth's generous, yet finite, boundaries.

Moreover, the global power arms race, leading to more dangerous weapons, is largely fueled by competition for dwindling resources. This fact makes the concept of a circular economy—a paradigm that decouples growth from resource plunder—relevant not only for liberal hippies who care about the environment but also for conservatives who care about national security. Your purchasing decisions can champion businesses that embrace circularity, be it through leasing instead of owning, buying secondhand, or opting for recycled materials. Better yet, before you commit to a purchase, ask yourself, *Will this truly make me happier in the long run?* Often you'll discover that a fulfilling life doesn't require an abundance of possessions.

Turning our gaze to the realm of investment, it's easy to underestimate the power at our disposal. Even if your portfolio of investments seems modest, remember that pension plans are a form of investment too. In Europe and North America, the average individual has $65,000 invested in pension funds. What good is a pension if there's no planet to enjoy it on? Your retirement savings are a lever of change. Challenge your fund managers to ensure that your investments reflect your values, fostering a world you wish to retire in.

For those wielding more substantial financial power—be they billionaires or investment firms managing billions—their influence can reshape industries. Engage in shareholder activism, make your voice heard at annual meetings, and drive the agenda toward sustainability and ethical practices. Some concrete things you can do as an investor include, but are not limited to:

- Divest from technologies and business models that pose risks to sustainability, safety, and the fulfillment of intrinsic values.
- Channel investments into start-ups within the circular economy sector, fostering growth that utilizes fewer resources and mitigates global conflicts and competition.
- Invest venture capital in start-ups dedicated to developing reputational markets, enhancing accountability, and ethical business practices.
- Leverage existing prediction markets as tools for risk assessment, identifying potential pitfalls in current and

prospective investments to spur market demand for these analytical tools.

- Mandate that your portfolio companies provide industry-relevant public disclosures reflecting intrinsic values, and push for transparency throughout their value chains. These measures will allow markets to better understand the risks and governance opportunities present.
- Advocate that ESG rating agencies incorporate these comprehensive disclosures into their evaluation criteria, promoting a more holistic approach toward assessing company performance and impact.

Perhaps the most potent move is either to spearhead or to influence investor coalitions. By pooling resources, investors can amplify their impact.

I recognize the limitations of the "vote with your wallet" approach as a mechanism for driving change. Its effectiveness is predominantly within the reach of the ultrawealthy—a subset of people who have not always been known for their keen interest in preventing corrosive selfishness—which highlights that this approach cannot replace the need for political and systemic actions. Instead, it should serve as a supplementary measure, existing alongside broader efforts to achieve meaningful impact.

Create Awareness

You've likely encountered the adage "The first step in solving a problem is acknowledging its existence." This adage holds particularly true for Darwinian demons. The more people know about these demons, the sooner we can step away from the "bad apples" narrative and start addressing root causes. Have you dealt with an unreasonable traffic citation, faced exorbitant medication prices at the pharmacy, been inundated with scams in your YouTube feed, or noticed politicians prioritizing victory over representation? Or has your rude, less capable colleague received an undeserved promotion?

If you observe people acting in a manner that seems antisocial, take a moment to consider the possible incentive structures influencing this behavior, and share your reflections with friends and family. Challenge those instances where the discourse devolves into "bad apples" finger-pointing, and instead try to encourage your interlocutors to address the root causes and underlying incentives with empathy and a deeper understanding. Commit to endorsing and supporting media outlets and journalism that adhere to these principles of thoughtful analysis and constructive dialogue. Actively participate in campaigns aimed at transforming the underlying incentives, particularly those posing threats to our planet.

If you're involved in journalism or advocacy, try to ensure that your revenue models reflect your core values. Dedicate more time and effort to promoting "demon literacy," educating the public about the underlying incentives and structures that may lead to potential existential threats. For academics in fields related to psychological inner development, game theory, economics, and policy, prioritize making your research accessible and impactful to the broader public.

If you, as an organization or individual, are taking positive steps to combat Darwinian demons, don't hesitate to share your achievements. Broadcast your efforts to the world and serve as an inspiration to others! But where do you start?

Go Demon Hunting

Having successfully established the habit of calling out Darwinian demons when you see them, the next step is to determine when and where it is in your power to try to defeat them. It will help to learn how to apply the three principles of value alignment within your immediate environment. Do you have the ability to shape the values of your workplace or local community (Principle 1)? Can you collaboratively brainstorm with your colleagues to identify actions that are likely to exert the greatest impact on these values (Principle 2)? Are you in a position to promote adherence to these behaviors (Principle 3)?

As a boss or manager, you certainly have the capacity to enact these principles in your local context. However, even as an individual contributor within a small team, you may still wield influence over your manager and fellow team members.

- **Lawyers:** Question the legality of behavior that might pose existential risk.
- **Insurance companies:** Use prediction markets to calculate premiums of risky behavior.
- **Entrepreneurs:** Race to build the first reputational market software.
- **Media:** Build business models that are aligned with your values of journalistic integrity.
- **Ratings agencies:** Consider the idea of reputational markets as a new and exciting revenue stream.

- **Enterprises:** Define your core values, and incorporate them into your performance and growth criteria, and bonus payment structures.

No matter where you work, most individuals have the opportunity to develop the sorts of social-emotional skills that can help them escape the pull of rational selfishness and instead consciously choose a more cooperative path. When we state our goals for the week, month, or year, how many of us also take the time to list our intrinsic values, then see how well the lists align? And how many of us brainstorm how our actions correspond with those values, and how we incentivize ourselves to follow through on them?

Understanding your core values and acting in a manner that amplifies them can profoundly impact your sense of purpose and well-being. This process of inner development is crucial not just for personal growth but also for cultivating a spirit of cooperation in the face of disagreements and competition. This is why even the UN Intergovernmental Panel on Climate Change recognizes the significance of inner development as a vital component in addressing climate change.

When you find yourself at odds with someone, ask yourself: *Is our disagreement rooted in fundamentally different values, or do we simply have diverse experiences and different ideas on how to realize those shared values?* In my experience, the latter is often the case. By focusing on your common values, you can engage in rational dialogue and employ empiricism to forge a path toward consensus on the most effective strategies to achieve your mutual goals. Most important, approach these discussions with empathy. Avoid the trap of labeling those you disagree with as malevolent. They may be operating within a system rife with perverse incentives, doing their best to navigate it; or they might come from a different set of experiences that lead to a unique perspective, all the while sharing your core values. For practical tools that can help you find your intrinsic values, and the most efficient ways to influence those values, I recommend the website ClearerThinking.org.

Why This Moment Matters

Every breathtaking vista that unfolds before us, every treasure we hold dear, every marvel of nature that infuses our world with wonder, everything we value—all these are the legacies of the battle between Darwinian demons and angels, where somehow the angels came out on top.

Yet the Fragility of Life Hypothesis tells us that the triumph of Darwinian angels is not a foregone conclusion. After what might be 4 billion years of cosmic happenstance, humanity could stand as the sole custodian of life within our local galaxy cluster, endowed with the unique ability to break free from the Darwinian shackles. Unlike the relentless spread of cancer, the virulence of viruses, or the merciless killing of predators, we possess the extraordinary potential to steer our destiny with reason, empathy, and foresight. By completing the Great Bootstrap, we safeguard not only ourselves but also countless generations without a voice to claim their right to existence.

Let us not view this responsibility with a sense of despair, but rather as an exciting call to action. Millennia from now, our descendants may look back with amusement at our early struggles to forge lasting cooperation, marveling that we once allowed Darwinian demons to dictate our fate. Yet they will also gaze upon this era with profound respect and gratitude, recognizing us as the pioneers who unlocked the formula for stable cooperation, who initiated our grand ascent. Our generation can mark the dawn of a cosmic odyssey, where humanity, armed with the gift of cooperation, embarks on a voyage to spread the light of life and beauty into the dark, cold, and vast cosmos.

GLOSSARY

Arms race: a competitive cycle of escalation where rivals continuously strive to outmatch each other's capabilities, often leading to rapid and sometimes unsustainable increases in resources spent on the area of competition. While such arms races are often net negative, the dramatic arms races we witness today—toward ever more resources, power, and intelligence—may imperil the very future of life on Earth.

Darwinian angel: a selection pressure that makes it adaptive for agents to positively impact others.

Darwinian demon: a selection pressure that makes it adaptive for agents to negatively impact others.

Evolution by natural selection: the process by which individuals better adapted to their environment tend to survive and produce more offspring, leading to the accumulation of favorable traits in a population over generations. Evolution by natural selection occurs in any domain in which there is variation (individual entities vary in their traits), selection (some traits are more conducive to survival and reproduction than others), and retention (those traits are to some extent passed on to the next generation).

Fragility of Life Hypothesis: the idea that life may be inherently fragile because evolution inevitably gives rise to Darwinian demons with the potential to obliterate all life.

Goodhart's Law: the observation that once a measure becomes a target for social or economic policy, it ceases to be a reliable indicator, because individuals will seek to manipulate or game the system in order to meet the target.

Great Bootstrap: a plan for defeating Darwinian demons once and for all, by replacing the short-sighted survival algorithm of natural selection with a new system. In this new framework, the primary objective is cooperation, centered on the fulfillment of our intrinsic values.

Indirect reciprocity: an evolutionary mechanism where individuals support others based on their reputation for helping, promoting cooperative behavior within a community without the need for direct reciprocation between individuals.

Major evolutionary transition: a significant shift in the organization and complexity of life, often involving the emergence of cooperation at higher levels, such as the transition from single-celled to multicelled organisms.

Multilevel selection: the idea that natural selection operates not only on individuals but also on groups, selecting traits that benefit the survival and reproduction of both individual organisms and the groups to which they belong. For example, under intense competition or warfare between different tribes, multilevel selection can favor the emergence of cooperative norms, since tribes with greater internal harmony are better equipped to fight others.

Narrow success metric: a measure that is particularly prone to manipulation via Goodhart's Law.

Reputational market: a speculative market where trade is centered on predictions about a company's future impact on the world, be it positive or negative.

Reputational score: a score assigned to companies and other entities on the basis of their predicted impact on the things we truly care about. By harnessing the power of indirect reciprocity, global adoption of a value-aligned reputational score could help bring about the Great Bootstrap.

Selection pressure: a feature of the environment that favors the survival or reproduction of certain traits or individuals over others.

Universal Darwinian drive: a type of trait that is adaptive under a broad range of selection pressures. There are universal Darwinian drives toward replication, repair, resources, power, and intelligence. Sometimes these drives are harnessed by Darwinian demons to create destructive outcomes, but they can equally be mobilized by Darwinian angels for the greater good.

NOTES

PROLOGUE
The Parable of Picher

ix **Picher's infant mortality rate:** Dan Shepherd, "Last Residents of Picher, Oklahoma Won't Give Up the Ghost (Town)," NBC News, April 26, 2014.

ix **The town was pockmarked:** Tom Lindley, "Who Will Save the Children of Ottawa County?" *Oklahoman,* December 9, 1999.

ix **And the people of Picher:** Shepherd, "Last Residents of Picher, Oklahoma Won't Give Up the Ghost (Town)."

ix **In the 1920s:** Dianna Everett, "Tri-State Lead and Zinc District," *The Encyclopedia of Oklahoma History and Culture,* https://www.okhistory.org/publications/enc/entry?entry=TR014.

x **The miners of Picher:** David Robertson, "Enduring Places: Landscape Meaning, Community Persistence, and Preservation in the Historic Mining Town," (PhD diss., University of Oklahoma, 2001), 208.

x **"There is probably no town":** "Converted Picher," *King Jack,* January 15, 1920, CardinKids.com.

x **as high as two hundred feet:** Ross Milloy, "Picher Journal; Waste from Old Mines Leaves Piles of Problems," *New York Times,* July 21, 2000.

x **By the 1960s:** Kenneth V. Luza and W. Ed Keheley, *Field Trip Guide to the Tar Creek Superfund Site, Picher, Oklahoma* (Oklahoma Geological Survey, 2006).

xi **a rate 2,000 percent higher:** Shepherd, "Last Residents of Picher, Oklahoma Won't Give Up the Ghost (Town)."

xi **38.3 percent of the children:** Lindley, "Who Will Save the Children of Ottawa County?"

xi **among the dead fish:** Richard E. Meyer, "Acid Water Drowns Tar Creek as Cleanup Delayed," *Oklahoman,* February 6, 1983.

xi **in the swimming holes:** Shepherd, "Last Residents of Picher, Oklahoma Won't Give Up the Ghost (Town)."

xi **In the early 1980s:** Luza and Keheley, *Field Trip Guide.*

xi **cash buyouts:** Shepherd, "Last Residents of Picher, Oklahoma Won't Give Up the Ghost (Town)."

xiii **The rat catchers of Hanoi:** Michael G. Vann, *The Great Hanoi Rat Hunt: Empire, Disease and Modernity in French Colonial Vietnam* (New York: Oxford University Press, 2018).

xvii **"Meditations on Moloch":** Scott Alexander, "Meditations on Moloch" (blogpost), SlateStarCodex.com, July 30, 2014.

CHAPTER ONE

A Crash Course on Darwinian Demons

5 **"As many more individuals":** Charles Darwin, *On the Origin of Species by Means of Natural Selection* (London: John Murray, 1859), 6.

6 **In 1983 the evolutionary biologist:** Richard Dawkins, "Universal Darwinism," in *Evolution from Molecules to Man,* ed. D. S. Bendall (Cambridge: Cambridge University Press, 1983).

11 **demon known as Spiegelman's Monster:** Richard Dawkins and Yan Wong, *The Ancestor's Tale: A Pilgrimage to the Dawn of Life,* 2nd ed. (London: Weidenfeld & Nicolson, 2016), chap. 39.

14 **The LAPD citation scandal:** Andrew Blankstein and Joel Rubin, "LAPD Officers Who Complained About Ticket Quotas Are Awarded $2 Million," *Los Angeles Times,* April 12, 2011; Joel Rubin and Catherine Saillant, "L.A. Approves $6-Million Settlement over Alleged Traffic Ticket Quotas," *Los Angeles Times,* December 10, 2013.

14 **Today citation quotas are:** Andy Corbley, "Virginia Joins 20 Other States Banning Ticket Quotas for Traffic Cops," GoodNewsNetwork.org, June 6, 2022.

17 **"Even when herdsmen":** Garrett Hardin, "The Tragedy of the Commons," *Science* 162, no. 3859 (1968): 1243–48.

18 **By the early 1990s:** Mark Kurlansky, *Cod: A Biography of the Fish That Changed the World* (London: Walker Books, 1997).

22 **"The total amount of suffering":** Richard Dawkins, *River Out of Eden: A Darwinian View of Life* (New York: Basic Books, 1995).

22 **"When a measure becomes a target":** David Manheim and Scott Garrabrant, "Categorizing Variants of Goodhart's Law," arXiv.org, February 24, 2019.

CHAPTER TWO
Everyday Demons

24 **"moments ranging from":** Sapna Maheshwari, "On YouTube Kids, Startling Videos Slip Past Filters," *New York Times,* November 4, 2017.

26 **Indeed, as the writer James Bridle:** James Bridle, "Something Is Wrong on the Internet," Medium.com, November 6, 2017.

26 **"My poor little innocent boy":** Maheshwari, "On YouTube Kids."

29 **"a micro-manager":** CMI Action, *Taking Responsibility: Why UK PLC Needs Better Managers,* Managers.org.uk, October 2023.

31 **blockbuster investigation:** E. Scott Reckard, "Wells Fargo's Pressure-Cooker Sales Culture Comes at a Cost," *Los Angeles Times,* December 21, 2013.

31 **"coercing customers":** Stacy Cowley, "Voices from Wells Fargo: 'I Thought I Was Having a Heart Attack,'" *New York Times,* October 20, 2016.

33 **"allows you to be":** "Boost: What Is Boost?" Help.Tinder.com, n.d.

35 **so-called beauty filters:** Liv Boeree, "The Moloch Trap of AI Beauty Filters" (video), n.d., at https://www.youtube.com/watch?v=fifVuhgvQQ8

35 **Gender and Sexualities Research Centre:** Rosalind Gill, *Changing the Perfect Picture: Smartphones, Social Media and Appearance Pressures* (London: University of London, 2021).

36 **"Recently a patient":** Erin Lukas, "The Most Popular Inspiration for Plastic Surgery Is Yourself," *InStyle,* June 23, 2021.

36 **"have particular difficulty":** Tate Ryan-Mosley, "Beauty Filters Are Changing The Way Young Girls See Themselves," *MIT Technology Review* April 2, 2021.

37 **customer satisfaction:** Sebastian Herrera, "Amazon's Customer Satisfaction Slips with Shoppers," *Wall Street Journal,* November 21, 2022.

38 **"Why does it feel":** John Herrman, "The Junkification of Amazon," *New York Magazine,* January 30, 2023.

39 **Food and Brand Lab:** James Hamblin, "A Credibility Crisis in Food Science," *Atlantic,* September 24, 2018; Peter Etchells and Chris Chambers, "Mindless Eating: Is There Something Rotten Behind the Research?" *Guardian,* February 16, 2018.

40 **A dedicated effort to reproduce studies:** C. F. Camerer et al., "Evaluating the Replicability of Social Science Experiments in Nature and Science Between 2010 and 2015," *Nature Human Behaviour* 2 (2018): 637–44.

40 **large international project:** R. A. Klein et al., "Many Labs 2: Investigating Variation in Replicability Across Sample and Setting," *OSF,* September 28, 2017.

40 **In the field of psychology:** Lukas, "Most Popular Inspiration"; Open Science

Collaboration, "Estimating the Reproducibility of Psychological Science," *Science* 349, no. 6251 (2015).

42 **"Members of Congress have become":** Michael Beckel, "'Why We Left Congress': Excerpts of Our Conversation with Rep. Rick Nolan (D-MN)," IssueOne.org, December 21, 2018.

45 **"a pattern of intimidation, threats":** "A Timeline of How the Atlanta School Cheating Scandal Unfolded," *Atlanta Journal-Constitution,* April 2, 2015.

47 **"We need to erect":** Barbara Matejčić, *In Veles, Meeting the Producers of Fake News* (European Trade Union Institute, 2018).

CHAPTER THREE
Angels, Demons, and the Epic Struggle of Life

51 **first documented case:** Steven I. Hajdu, "A Note From History: Landmarks in History of Cancer, Part 1," *Cancer* 117, no. 5 (2011): 1097–102.

52 **Virchow's research helped:** Edward Walter and Mike Scott, "The Life and Work of Rudolf Virchow 1821–1902: 'Cell Theory, Thrombosis and the Sausage Duel,'" *Journal of the Intensive Care Society* 18, no. 3 (2017): 234–35.

57 **What may be perceived as:** C. Athena Aktipis et al., "Cancer Across the Tree of Life: Cooperation and Cheating in Multicellularity," *Philosophical Transactions of the Royal Society B: Biological Sciences* 370 (2015).

59 **the suppression of cheating:** Andrew H. Knoll, "The Multiple Origins of Complex Multicellularity," *Annual Review of Earth and Planetary Sciences* 39 (2011): 217–39.

59 **"There can be no doubt":** Charles Darwin, *The Descent of Man, and Selection in Relation to Sex* (London: John Murray, 1871), 166.

60 **Indeed, the creation of rule-based:** David Sloan Wilson, Mark Van Vugt, and Rick O'Gorman, "Multilevel Selection Theory and Major Evolutionary Transitions: Implications for Psychological Science," *Current Directions in Psychological Science* 17, no. 1 (2008): 6–9; David Sloan Wilson et al., "Multilevel Cultural Evolution: From New Theory to Practical Applications," *Proceedings of the National Academy of Sciences USA* 120, no. 16 (2023); Peter J. Richerson and Robert Boyd, *Not By Genes Alone: How Culture Transformed Human Evolution* (Chicago: University of Chicago Press, 2005).

64 **Asphaltization:** Marc Neveu, Hyo-Joong Kim, and Steven A Benner, "The 'Strong' RNA World Hypothesis: Fifty Years Old," *Astrobiology* 13, no. 4 (April 2013): 391–403.

64 **Reverse Krebs Cycle:** R. Trent Stubbs, Mahipal Yadav, Ramanarayanan Krishnamurthy, and Greg Springsteen, "A Plausible Metal-Free Ancestral Analogue

of the Krebs Cycle Composed Entirely of α-ketoacids," *Nature Chemistry* 12 (2020): 1016–22.

64 **Selfish genetic elements:** Leslie A. Pray, "Transposons: The Jumping Genes," *Nature Education* 1, no. 1 (2008): 204.

64 **150 million:** Barbara L. Thorne, David A. Grimaldi, and Kumar Krishna, "Early Fossil History of the Termites," in *Termites: Evolution, Sociality, Symbioses, Ecology,* ed. Tayuka Abe, David E. Bignell, and Masahiko Higashi (Dordrecht: Springer, 2000).

64 **Eusociality:** Martin A. Nowak, Corina E. Tarnita, and Edward O .Wilson, "The Evolution of Eusociality," *Nature* 466, no. 7310 (2018): 1057–62.

64 **Language:** Robert C. Berwick and Noam Chomsky, *Why Only Us: Language and Evolution* (Cambridge, MA: MIT Press, 2017).

CHAPTER FOUR
Demons That Threaten Life Itself

68 ***kamikaze mutants*:** Hiroyuki Matsuda and Peter A. Abrams, "Timid Consumers: Self-Extinction Due to Adaptive Change in Foraging and Anti-Predator Effort," *Theoretical Population Biology* 45, no. 1 (1994): 76–91; Hiroyuki Matsuda and Peter A. Abrams, "Runaway Evolution to Self-Extinction Under Asymmetrical Competition," *Evolution* 48, no. 6 (1994): 1764–72.

68 **Why do scientists keep finding:** Colleen Webb and Joseph Travis, "A Complete Classification of Darwinian Extinction in Ecological Interactions," *American Naturalist* 161, no. 2 (2003); Christoph Ratzke, Jonas Denk, and Jeff Gore, "Ecological Suicide in Microbes," *Nature Ecology and Evolution* 2 (2018): 867-72; Michael E. Gilpin, "Prudent Predation and the Character of Ecological Attractor Sets," *Rocky Mountain Journal of Mathematics* 9, no. 1 (1979): 83–86.

71 **the bullet holes found in:** Abraham Wald, *A Method of Estimating Plane Vulnerability Based on Damage of Survivors* (Statistical Research Group, Columbia University, 1943).

73 **an urn filled with balls:** Nick Bostrom, "The Vulnerable World Hypothesis," *Global Policy* 10, no. 4 (2019).

77 **The paleontologist Peter Ward:** Peter Ward, *The Medea Hypothesis: Is Life on Earth Ultimately Self-Destructive?* (Princeton, NJ: Princeton University Press, 2009).

77 **Great Oxidation Event:** Malcolm S. W. Hodgskiss et al., "A Productivity Collapse to End Earth's Great Oxidation," *Proceedings of the National Academy of Sciences USA* 116, no. 35 (2019): 17207–12.

78 **Sturtian and Marinoan "Snowball Earth" Glaciations:** R. J. Stern, D. Avigad, N. R. Miller, and M. Beyth, "Evidence for the Snowball Earth Hypothesis in the Arabian-Nubian Shield and the East African Orogen," *Journal of African Earth Sciences* 44, no. 1 (2006): 1–20. Alan D. Rooney, Justin V. Strauss, Alan D. Brandon, and Francis A. Macdonald, "A Cryogenian Chronology: Two Long-Lasting Synchronous Neoproterozoic Glaciations," *Geology* 43, no. 5 (2015): 459–62.

78 **Late Ordovician Mass Extinction:** Timothy M. Lenton et al., "First Plants Cooled the Ordovician," *Nature Geoscience* 5 (2012): 86–89.

78 **Late Devonian Extinction:** T. J. Algeo, "Terrestrial-Marine Teleconnections in the Devonian: Links Between the Evolution of Land Plants, Weathering Processes, and Marine Anoxic Events," *Philosophical Transactions of the Royal Society B: Biological Sciences* 353, no. 1365 (1998): 113–30.

78 **Permian-Triassic Extinction:** Changqun Cao et al., "Biogeochemical Evidence for Euxinic Oceans and Ecological Disturbance Presaging the End-Permian Mass Extinction Event," *Earth and Planetary Science Letters* 281, nos. 3–4 (2009): 188–201.

78 **Triassic-Jurassic Extinction:** Sylvain Richoz et al., "Hydrogen Sulphide Poisoning of Shallow Seas Following the End-Triassic Extinction," *Nature Geoscience* 5 (2012): 662–67.

80 **an intelligent civilization:** Stuart Armstrong and Anders Sandberg, "Eternity in Six Hours: Intergalactic Spreading of Intelligent Life and Sharpening the Fermi Paradox," *Acta Astronautica* 89 (2013): 1–13.

80 **major evolutionary transitions:** Andrew E. Snyder-Beattie, Anders Sandberg, K. Eric Drexler, and Michael B. Bonsall, "The Timing of Evolutionary Transitions Suggests Intelligent Life Is Rare," *Astrobiology* 21, no. 3 (2021).

81 **These scenarios aren't merely speculative:** Toby Ord, *The Precipice: Existential Risk and the Future of Humanity* (London: Bloomsbury, 2020); Will MacAskill, *What We Owe The Future* (New York: Basic Books, 2022).

81 **Voluntary Human Extinction:** Voluntary Human Extinction Movement, https://www.vhemt.org/.

CHAPTER FIVE

The Depleted World: The Resources Arms Race

88 **improved technologies for forest harvesting:** Hannah Ritchie, "The World Has Lost One-Third of Its Forest, But an End of Deforestation Is Possible," OurWorldinData.org, February 9, 2021.

93 **Current estimates suggest:** International Energy Agency, *World Energy Outlook 2023,* iea.org, October 24, 2023.

98 **nine recognized planetary boundaries:** Johan Rockström et al., "A Safe Operating Space for Humanity," *Nature* 461 (2009): 472–75.

98 **Our global thermostat:** Gail Hartfield, Jessica Blunden, and Derek S. Arndt, *State of the Climate in 2017* (American Meteorological Society, 2018).

98 **The rate of species extinction:** Christopher N. Johnson, "Past and Future Decline and Extinction of Species," Royal Society.org, n.d.

98 **the phosphorus and nitrogen cycles:** K. Richardson et al., "Earth Beyond Six of Nine Planetary Boundaries," *Science Advances* 9, no. 37 (2023).

101 **many have found ways to decouple:** Hannah Ritchie, "Many Countries Have Decoupled Economic Growth from CO2 Emissions, Even If We Take Offshored Production into Account," OurWorldinData.org, 2021.

103 **Yet between 1998 and 2014:** Geoffrey Supran, Stefan Rahmstorf, and Naomi Oreskes, "Assessing ExxonMobil's Global Warming Projections," *Science* 379, no. 6628 (2017); Chloe Farand, "ExxonMobil Talks Up Tackling Climate Change 'While Still Funding Climate Deniers,'" *Independent,* July 1, 2017.

104 **working paper titled "Emissions Gaming?":** Marc Lepere et al., "Emissions Gaming?," *Science Advances* 9, no. 37 (2023).

104 **recent study, involving twenty-six projects:** Thales A. P. West et al., "Action needed to make carbon offsets from forest conservation work for climate change mitigation," *Science* 381, 873-77 (2023).

107 **each standard deviation increase:** Solomon M. Hsiang, Marshall Burke, and Edward Miguel, "Quantifying the Influence of Climate on Human Conflict," *Science* 341, no. 6151 (2013). Sida, *The Relationship Between Climate Change and Violent Conflict,* CDN.Sida.se, 2017).

CHAPTER SIX

Extinction Weapons: The Power Arms Race

109 **Within two hundred years:** Alexander Koch, Chris Brierley, Mark Maslin, and Simon Lewis, "European Colonization of the Americas Killed 10 Percent of World Population and Caused Global Cooling," TheWorld.org, January 31, 2019.

111 ***offensive realism*:** John Mearsheimer, *The Tragedy of Great Power Politics* (New York: W.W. Norton, 2001).

112 **Consider the Mongol Empire:** Timothy May, *Mongol Art of War: Chinggis Khan and the Mongol Military System* (Westholme, 2007).

113 **Napoleon Bonaparte:** Philip J. Haythornthwaite, *Napoleon's Military Machine* (Cambridge, MA: Da Capo Press, 1995).

115 **physicist J. Robert Oppenheimer:** Kai Bird and Martin J. Sherwin, *American*

Prometheus: The Triumph and Tragedy of J. Robert Oppenheimer (New York: Vintage Books, 2006).

115 **Even a "small" regional nuclear war:** Alan Robock, Georgiy Stenchikov, O. B. Toon, and L. Oman, "Climatic Consequences of Regional Nuclear Conflicts," *Atmospheric Chemistry and Physics* 7 (2007): 2003–12.

116 **more than thirteen thousand warheads:** Hans Kristensen, Matt Korda, Eliana Johns, and Kate Kohn, *Status of World Nuclear Forces* (Federation of American Scientists, 2023).

116 **direct and indirect death toll:** Lili Xia et al., "Global Food Insecurity and Famine from Reduced Crop, Marine Fishery and Livestock Production Due to Climate Disruption from Nuclear War Soot Injection," *Nature Food* 3 (2022): 586–96.

118 **Goldsboro Incident:** "First Things First: It Did Happen," *The Full Story,* https://web.archive.org/web/20050306185714fw_/http://www.ibiblio.org/bomb/initial.html.

118 **A 1969 official report:** Parker F. Jones, "Goldsboro Revisited, or How I Learned to Mistrust the H-Bomb, or To Set the Record Straight," (Albuquerque: Sandia Laboratories, October 22, 1969).

119 **"a Sunday picnic":** "At the Heart of C3I Are Computers. The Biggest . . ." UPI, August 2, 1987.

119 **"Cocked Pistol":** James Coates, "War Games: The Computer Is a General," *Chicago Tribune,* May 1, 1983.

120 **"was going to war":** Daniel Uria, "False Alarm: 1979 NORAD Scare Was One of Several Nuclear Close Calls," *UPI*, November 8, 2019.

120 **a B-52 bomber took off:** US Air Force, "An Unauthorized Transfer of Nuclear Warheads Between Minot AFB, North Dakota and Barksdale AFB, Louisiana, 30 August 2007" (document), https://scholar.harvard.edu/files/jvaynman/files/minot_afb_report.pdf

121 **previously unknown incidents:** Eric Schlosser, *Command and Control: Nuclear Weapons, the Damascus Accident, and the Illusion of Safety* (New York: Penguin Books, 2013).

125 **The cost of sequencing:** National Human Genome Research Institute, "The Cost of Sequencing a Human Genome," Genome.gov.

125 **Many experts on extinction:** Toby Ord, *The Precipice: Existential Risk and the Future of Humanity* (London: Bloomsbury, 2020), 167.

126 **pandemic forecasting firm:** Eleni Smitham and Amanda Glassman, "The Next Pandemic Could Come Soon and Be Deadlier," Center for Global Development, August 25, 2021.

126 **about 0.9 percent:** Anthony M. Barrett, Seth D. Baum, and Kelly Hostetler,

"Analyzing and Reducing the Risks of Inadvertent Nuclear War Between the United States and Russia," *Science and Global Security* 21(2013): 106–33.

CHAPTER SEVEN

Godlike AI: The Intelligence Arms Race

128 **forty thousand weaponizable toxic molecules:** Fabio Urbina, Filippa Lentzos, Cédric Invernizzi, and Sean Ekins, "Dual Use of Artificial-Intelligence-Powered Drug Discovery," *Nature Machine Intelligence* 4 (2022): 189–91.

128 **"It just felt a little surreal":** Rebecca Sohn, "AI Drug Discovery Systems Might Be Repurposed to Make Chemical Weapons, Researchers Warn," *Scientific American,* April 21, 2022.

129 **an estimated $154 billion:** Leigh McGowran, "Global AI Spending Expected to Hit $154bn This Year," SiliconRepublic.com, March 13, 2023.

131 **"intelligence measures":** Shane Legg and Marcus Hutter, "Universal Intelligence: A Definition of Machine Intelligence," *Minds and Machines* 17 (2007): 391–444.

132 **proposed a link between:** Peter J. Richerson and Robert Boyd, "Climate, Culture, and the Evolution of Cognition," in *Evolution of Cognition,* ed. Cecilia Heyes and Ludwig Huber, 329–46 (Cambridge, MA: MIT Press, 2000).

134 **"The single most important":** US Senate, Committee on the Judiciary, "Oversight of A.I.: Principles for Regulation" (video), Judiciary.Senate.gov, July 25, 2023.

134 **from 1952 to 2018:** Epoch, "Machine Learning Trends," Epochai.org, 2023.

134 **In comparison, the difference:** Suzana Herculano-Houzel, "The Remarkable, Yet Not Extraordinary, Human Brain as a Scaled-Up Primate Brain and Its Associated Cost," *Proceedings of the National Academy of Sciences USA* 109, supp. 1 (2012): 10661–68.

136 **the age of *foundation models*:** Rishi Bommasani et al., "On the Opportunities and Risks of Foundation Models," arXiv.org, July 12, 2022.

138 **"a moody, manic-depressive":** Kevin Roose, "A Conversation with Bing's Chatbot Left Me Deeply Unsettled," *The New York Times,* February 17, 2023.

139 **to do with *misspecification*:** Dario Amodei et al., "Concrete Problems in AI Safety," arXiv.org, July 25, 2016.

139 **In a 2017 paper:** Ivaylo Popov et al. "Data-efficient Deep Reinforcement Learning for Dexterous Manipulation," arXiv.org, April 10, 2017.

141 ***treacherous turn*:** Nick Bostrom, *Superintelligence: Paths, Dangers, Strategies* (New York: Oxford University Press, 2014).

141 **"In the hypothetical misgeneralised":** Rohin Shah et al., "Goal Misgeneralisation: Why Correct Specifications Aren't Enough For Correct Goals," arXiv.org, November 2, 2022.

142 **Marvin Minsky once speculated:** Stuart J. Russell and Peter Norvig, *Artificial Intelligence: A Modern Approach,* 2nd ed. (Englewood Cliffs, NJ: Prentice Hall, 2003), sec. 26.3.

143 **autonomous vehicles:** Jean-François Bonnefon et al., "The Social Dilemma of Autonomous Vehicles," *Science* 352, no. 6293 (2016): 1573–76.

144 **several proofs:** Mario Brčić and Roman V. Yampolskiy, "Impossibility Results in AI: A Survey," *ACM Computing Surveys* 56, no. 1 (2023): 1–24.

145 **Nick Bostrom's famous thought experiment:** Bostrom, *Superintelligence,* 123.

146 **"The AI doesn't love you":** Eliezer Yudkowsky, "Artificial Intelligence as a Positive and Negative Factor in Global Risk," in *Global Catastrophic Risks,* ed. Nick Bostrom and Milan M. Ćirković, 308–45 (New York: Oxford University Press, 2008).

147 **"Every eighteen months":** Eliezer Yudkowsky, "Three Major Singularity Schools" (video), talk at Singularity Summit, 2007, https://www.youtube.com/watch?v=mEt1Wfl1jvo&t=1230s

149 **AI safety researcher Dan Hendrycks:** Dan Hendrycks, "Natural Selection Favors AI over Humans," arXiv.org, July 18, 2023.

152 **"Mitigating the risk":** Center for AI Safety, "Statement on AI Risk," Safe.ai.

153 **"all AI labs to immediately":** Future of Life Institute, "Pause Giant AI Experiments: An Open Letter," FutureofLife.org, March 22, 2023.

155 **Sam Altman, the CEO:** Max Chafkin and Rachel Metz, "What We Know So Far About Why OpenAI Fired Sam Altman," *Time,* November 20, 2023.

155 **Musk alleges that OpenAI:** Dan Milmo, "Elon Musk Sues OpenAI Accusing It of Putting Profit Above Humanity," *Guardian,* March 1, 2024.

CHAPTER EIGHT
The Last Transition

166 **outlined the five mechanisms:** Martin Nowak, "Five Rules for the Evolution of Cooperation," *Science* 314, no. 5805 (2006): 1560–63.

CHAPTER NINE
Centralized Solutions Based on Force

171 **"We must embrace international cooperation":** Bernard Baruch, speech to the first session of the United Nations Atomic Energy Commission, New York, NY, June 14, 1946.

179 **"In 2016 the British economy":** Edward Luce, "Mark Carney: 'Doubling Down on Inequality Was a Surprising Choice,'" *Financial Times,* October 14, 2022.

179 **By uniting its member states:** Martin Dedman, *The Origins and Development of the European Union, 1945–2008: A History of European Integration,* 2nd ed. (New York: Routledge, 2010); Alasdair Blair, *The European Union Since 1945* (New York: Pearson, 2005).

181 **"Whoever becomes the leader in":** Ian Sample, "'Human Compatible' by Stuart Russell Review—AI and Our Future," *Guardian,* October 24, 2019.

182 **multinational artificial general:** Jason Hausenloy, Andrea Miotti, and Claire Dennis, "Multinational AGI Consortium (MAGIC): A Proposal for International Coordination on AI," arXiv.org, October 13, 2023.

183 **Bletchley Declaration:** UK Government, "The Bletchley Declaration by Countries Attending the AI Safety Summit," www.gov.uk, November 1–2, 2023.

185 **own the supply chains:** Saif M. Khan, Dahlia Peterson, and Alexander Mann, "The Semiconductor Supply Chain: Assessing National Competitiveness," Center for Security and Merging Technology, Georgetown University, January 2021.

185 **Introducing advanced monitoring:** Yonadav Shavit, "What Does It Take to Catch a Chinchilla? Verifying Rules on Large-Scale Neural Network Training via Compute Monitoring," arXiv.org, May 30, 2023.

185 **"know-your-customer" protocols:** Janet Egan and Lennart Heim, "Oversight for Frontier AI Through a Know-Your-Customer Scheme for Compute Providers," arXiv.org, October 20, 2023.

186 **"The Vulnerable World Hypothesis":** Nick Bostrom, "The Vulnerable World Hypothesis," *Global Policy* 10, no. 4, (November 2019): 455–76.

CHAPTER TEN

Indirect Reciprocity and the Power of Reputation

196 **logjam caused a massive disruption:** Vivian Yee and James Glanz, "How One of the World's Biggest Ships Jammed the Suez Canal," *The New York Times,* July 17, 2021.

197 **About fifteen years ago:** Lauren Gill and Daniel Moritz-Rabson, "Companies Already Ban the Use of Their Drugs for Lethal Injection. Now They're Blocking IV Equipment," *Intercept,* September 14, 2023.

199 **likened to rare-earth metals:** Erin Griffith, "The Desperate Hunt for the A.I. Boom's Most Indispensable Prize," *New York Times,* August 16, 2023.

200 **electronic components:** Kristina Partsinevelos and Cait Freda, "The Chip

Industry's Open Secret: Adversaries' Military Tech Relies on US Components," CNBC, April 17, 2023.

202 **A staggering 90 percent:** Principles for Responsible Investment, *2022–23 Annual Report,* UN Environment Program and UN Global Compact (2023).

202 **average correlation in ratings:** Florian Berg, Julian F. Kölbel, and Roberto Rigobon, "Aggregate Confusion: The Divergence of ESG Ratings," *Review of Finance* 26, no. 6 (2022): 1315–44.

203 **ESG ratings can be gamed:** Juan Destribats, "Tesla Finds Itself Trailing Behind Shell on ESG Scores, Why?" *Motor Finance Online,* October 13, 2023.

CHAPTER ELEVEN
Measuring What Matters

206 **expansive needs for infrastructure:** Lance Elliot LaGroue, "Accounting and Auditing in Roman Society," PhD diss., University of North Carolina at Chapel Hill, 2014.

207 **Florentine merchants:** Jane Gleeson-White, *Double Entry: How the Merchants of Venice Created Modern Finance* (New York: W.W. Norton, 2012).

211 **Jeff Bezos's acquisition:** Elahe Izadi and Will Sommer, "Washington Post Cuts Follow Rapid Expansion, Unmet Revenue Projections," *Washington Post,* October 11, 2023.

213 **"intrinsic values test":** Spencer Greenberg and Amber Dawn Ace, "Valuism: Doing What You Value as a Life Philosophy," ClearerThinking.org, March 30, 2023.

219 **"superforecasters":** Philip E. Tetlock and Dan Gardner, *Superforecasting: The Art and Science of Prediction* (New York: Crown, 2015).

222 **This practice, in turn, could distort:** V. Lizka, "Issues with Futarchy," RethinkPriorities.org, October 7, 2021.

225 **In the OECD countries alone:** Global Pension Statistics Project, "Pension Assets Fell by 14% in 2022, Down to USD 51 Trillion in the OECD," Organisation for Economic Co-operation and Development, November 12, 2023.

228 **"antidote technologies":** Vitalik Buterin, "My Techno-Optimism," Vitalik .eth.limo, November 27, 2023.

228 **Far-UVC irradiation:** Ernest R. Blatchley III et al., "Far UV-C Radiation: An Emerging Tool for Pandemic Control," *Critical Reviews in Environmental Science and Technology* 53, no. 6 (2023).

228 **advanced wastewater surveillance:** Megan B. Diamond et al., "Wastewater Surveillance of Pathogens Can Inform Public Health Responses," *Nature Medicine* 28 (2022): 1992–95.

228 **"differential technology development":** Jonas Sandbrink et al., "Differential

Technology Development: An Innovation Governance Consideration for Navigating Technology Risks," Papers.SSRN.com, December 3, 2023.

229 **The Turkey Problem:** Nassim Nicholas Taleb, *The Black Swan: The Impact of the Highly Improbable* (New York: Random House, 2007).

CHAPTER TWELVE
What You Can Do

240 **"conspicuous acts of kindness":** Lex Fridman, "Elon Musk: War, AI, Aliens, Politics, Physics, Video Games, and Humanity" (video), *Lex Fridman Podcast,* November 9, 2023.

246 **even the UN Intergovernmental Panel:** Fatima Denton et al., "Accelerating the Transition in the Context of Sustainable Development," in *Report of the Intergovernmental Panel on Climate Change,* ed. P. R. Shukla et al. (New York: Cambridge University Press, 2022).

ACKNOWLEDGMENTS

Writing this book has been a journey of exploration, reflection, and immense learning. It would not have been possible without the collective effort and support of many individuals for whom I am deeply grateful.

At the very heart of this journey is my core team, whose support was indispensable. Despite my name being the only one on the cover, the truth is that producing this book was genuinely a collaborative endeavor. Justin Peters, your meticulous efforts in manuscript development and preparation were invaluable, elevating the quality of this book and truly taking it to the next level. Madeleine Ahlström, your exceptional project management, intelligent input, and keen attention to detail has been a critical part of our success. Aron Vallinder, your rigorous research and analytical prowess significantly enriched the depth of this project. Veronica Rönn, your stunning illustrations breathed life into my concepts, beautifully visualizing the essence of our work. The contributions each of you made were foundational to the creation of this book, and for that, I am deeply grateful.

My sincere appreciation extends to Penguin Random House for embracing this project. Paul Whitlatch and Katie Berry, your faith in the potential of this book and your editorial guidance have been instrumental in shaping it into what it is today. Your belief and input have been a cornerstone of this endeavor. At Penguin Random House, I am also grateful to Janet Biehl, who provided a vigorous and valuable copyedit; eagle-eyed production editor Terry Deal; managing editors Christine Tanigawa and Liza Stepanovich; publicists Stacey Stein and Dyana Messina; marketers Chantelle Walker and Julie Cepler; and the executives of the Crown Publishing Group at Penguin Random House: David Drake, Gillian Blake, and Annsley Rosner. This book's striking cover was designed by Yang Kim, and its sleek interior page layouts were designed by Aubrey Khan. My profound thanks are due to all of these people, as well as the many others who worked behind the scenes to help bring this book to life.

I am deeply grateful to a distinguished group of intellectuals whose exceptional insights into the complexities of multipolar traps have profoundly influenced my understanding. I extend my heartfelt thanks to Daniel Schmachtenberger, Liv Boeree, Chris Anderson, Rob Reid, and Justin Shovelain. Your engaging discussions and perspectives have been a beacon of inspiration, guiding my journey. Your contributions are of immense value, and I sincerely appreciate the impactful work you continue to do. A special note of appreciation goes to Siri Helle for her astute and thoughtful feedback on the initial draft of our manuscript. The richness and insightfulness of your input have significantly enhanced the quality of this work, and for that, I am extremely grateful. Furthermore, I would like to acknowledge Scott Alexander. Though we have never met in person, his pivotal blog post on Moloch, penned more than a decade ago, was the catalyst for the founding of my company. The principles outlined in that post are intricately interwoven throughout the pages of this book, serving as a foundational element of its content.

I would like to express my sincere appreciation to my former colleagues at the Future of Humanity Institute (FHI) and the Centre for Effective Altruism at the University of Oxford. Their influence has been pivotal in shaping my perspectives on risk and the future. My deepest thanks go to Anders Sandberg, Toby Ord, Eric Drexler, Carl Frey, Will MacAskill, and Robert Wiblin for shaping my thinking. A special acknowledgment is reserved for Seán Ó hÉigeartaigh and Nick Bostrom, who not only enriched my intellectual journey with their groundbreaking work but also welcomed me into the FHI. Joining your ranks more than a decade ago marked a defining moment in my life, refining my thinking and bolstering my belief in the value of my ideas. Furthermore, I am deeply indebted to Max Tegmark, whose supportive email was a critical factor in my decision to move to Oxford as a young student.

I also want to thank all of my friends from the climate community who have taught me how to bridge academic theory and practical action. My gratitude extends to Johan Rockström and Johan Falk, who illuminated the delicate balance of Earth's systems; Kaya Axelsson, who artfully demonstrated how to transform boring standards work into exciting tools for climate impact; David Marriage, who showed the significance of clear demand signals across global value-chains to achieve net-zero; and Heather Buchanan, who unveiled the power of finance and banking in incentivizing business decarbonization from the top down.

To my esteemed cofounders, our extraordinary C-suite and leadership team, our invaluable investors, committed customers, and every dedicated colleague at Normative, your unwavering commitment to our shared vision has been the foundation of my drive and optimism. Our achievements in value-chain engagement and the pioneering of the Carbon Network have instilled in me a renewed hope, affirming my belief that we can indeed mitigate climate risk and, more broadly, global catastrophic risks through strategic value-chain governance. Your relentless support and encouragement have been crucial in driving our company's progress, significantly enhancing our mission's impact and demonstrating the feasibility and practicality of the ideas presented in this book.

Last but certainly not least, my deepest gratitude goes to my family. To my parents, who instilled in me the values of independent thinking and resilience. Your unconditional support in every decision I've made, from adopting vegetarianism at a tender age to venturing into entrepreneurship, has been my backbone. Your love and belief in me have been my guiding light!

To all mentioned and unmentioned who have contributed to this journey, your impact has been profound. Thank you for being part of this significant chapter in my life.

ART CREDITS

All images drawn or redrawn by Veronica Rönn.

page 44: US Supreme Court, 517 U.S. 952 (1996).

page 71: CC BY-SA 4.0 DEED.

page 76: CC BY 4.0 DEED.

page 95: Redrawn with data from the UK Ministry of Defence, *Global Strategic Trends—The Future Starts Today,* 6th ed., Assets.Publishing.Service.Gov.UK, October 2, 2018.

page 99: Redrawn with data from Hannah Ritchie, "Wild Mammals Make Up Only a Few Percent of the World's Mammals," *Our World in Data,* December 15, 2022 (CC BY).

page 100: Redrawn with data from Stockholm Resilience Centre, based on analysis in K. Richardson et al., "Earth Beyond Six of Nine Planetary Boundaries," *Science Advances* 9, no. 37 (2023) (CC BY-NC-ND 3.0).

page 122: Redrawn with data from Seth D. Baum, Robert de Neufville, and Anthony M. Barrett, "A Model for the Probability of Nuclear War," Global Catastrophic Risk Institute, Working Paper no. 18-1, March 8, 2018.

page 135: Redrawn with data from Will Henshall, "Four Charts That Show Why AI Progress Is Unlikely to Slow Down," *Time,* August 2, 2023.

page 137: Redrawn with data from Jaime Sevilla et al., "Compute Trends Across Three Eras of Machine Learning," arXiv.org, March 9, 2022.

page 142: Redrawn with data from Rohin Shah et al., "Goal Misgeneralisation: Why Correct Specifications Aren't Enough for Correct Goals," *Medium,* October 7, 2022.

page 198: Redrawn based on DarwinPeacock, Maklaan, Social Network Diagram (segment).svg, Wikimedia, CC BY 3.0 DEED.

page 209: Redrawn with data from BDO USA, LLP © 2021 BDO USA, LLP. All rights reserved. https://blogs.cfainstitute.org/investor/2021/08/10/esg-ratings-navigating-through-the-haze/.

page 229: Redrawn with data from Kevin McCordic, "1001 Days: How to Beat the Turkey Problem (And How It Applies to Bitcoin)," *Medium,* October 17, 2019.

INDEX

ABOUT THE AUTHOR

KRISTIAN RÖNN is the CEO and cofounder of Normative, a tech company that automates companies' carbon accounting. He has a background in mathematics, philosophy, computer science, and artificial intelligence.

Before he started Normative, he worked at the University of Oxford's Future of Humanity Institute on issues related to global catastrophic risks. Kristian is a thought leader within carbon accounting, with speaking engagements at COP, Davos, and Stockholm+50, as well as appearances in media outlets such as Bloomberg and Sky News. He has advised governments and international bodies, and has been officially acknowledged by the UNDP for his contribution to Global Sustainability Goal number 13: Stop Climate Change.